Seeds of
BLOOD and BEAUTY

Seeds of BLOOD and BEAUTY

SCOTTISH PLANT EXPLORERS

Ann Lindsay

BIRLINN

This revised and corrected edition first published in 2008 by
BIRLINN LIMITED
West Newington House
10 Newington Road
Edinburgh EH9 1QS
www.birlinn.co.uk

ISBN 13: 978 1 84158 579 6
ISBN 10: 1 84158 579 3

British Library Cataloguing-in-Publication Data
A catalogue record for this book is available from the British Library

Design: Mark Blackadder

Typeset by Brinnoven, Livingston
Printed and bound by CPI Cox & Wyman, Reading, RG1 8EX

Contents

Acknowledgements

Among the many who helped in the research and writing of this book, I would like to thank: my son, Robert Bradley, who reduced to one digestible chapter the biography of David Douglas, written by myself and Syd House; Patricia Waite, who lent volumes of *Familiar Garden Flowers* by Shirley Hibberd/ F. Edward Hulme, from which many illustrations have been used within the book; Jack Still, Jenny Thomson of Auchinblae and Jane Cruickshank at Mearns Leader who contributed to the chapter on David Lyall; Fiona and Louise Meikle, descendants of the Drummond family, who contributed greatly to information about Thomas Drummond; Julia Corden, Garden Manager at the Explorers Garden, Pitlochry, Perthshire, and her predecessor there, Nick Dawson; and Professor Stephen Blackmore, Director of the Royal Botanic Garden, Edinburgh.

Librarians are the essential but background stars for anyone undertaking copious research, and among those I need especially to thank are: Gina Douglas, Librarian & Archivist, Linnean Society of London; Liz Gilbert at the Lindley Library, Royal Horticultural Society, London; Leander Wolstenholme at the Department of Botany at the Liverpool Museum; Leonie Paterson, Ruth Hourston, Graham Hardy and Jane Hutcheon, the Library of the Royal Botanic Garden, Edinburgh; Karen Hartnup, University of Edinburgh special collections; Jan Morison of the Local History Department at Aberdeen Library; Jeremy Duncan of the Local History Department at the A.K. Bell Library, Perth; and librarians at the Forfar, Montrose and Inverness Libraries.

Ann Lindsay
July 2008

Introduction

A handful of Scots, roughly the same in number as a smallish class in a school or the Scottish football or rugby team with a few reserves on the side, have hunted out and introduced into the West more plants from around the world than probably all the other European nations combined. Every garden in Britain, and those of most of our European neighbours as well, contains plants originally brought back to Europe by a Scot. From the early eighteenth century onwards, Scots' expertise, cunning, curiosity, intelligence, adventurousness of spirit, scientific knowledge and business acumen dominated the horticultural scene not only in Great Britain, but around the world. These Scottish plant experts were as influential as Scottish engineers, doctors and businessmen. Yet theirs remains one of the great untold tales. This book reveals their story.

From the early eighteenth century onwards, many young Scotsmen left their native shores to seek their fortunes elsewhere. For young Scottish gardeners opportunities came via their older counterparts who had already established themselves in the service of wealthy aristocrats who appreciated their work ethic, detailed knowledge and sheer craftsmanship. These Scottish head gardeners south of the Border were able to set up a sort of bridgehead, enabling young Scotsmen of promise to join them as under gardeners. Other keen young gardeners worked for the Royal Botanic Garden in Edinburgh, which became a great source of likely plant explorers. Yet another source of plant collectors was found amongst the ranks of young Scottish surgeons who, on graduating from the Scottish medical institutions, found their most likely method of advancement was by joining the Royal Navy. Their rigorous medical training, which usually included copious study of botany, opened up further career possibilities.

From such varied backgrounds came young men who saw plant collecting in little-known foreign lands as a rare and golden opportunity to burst out of the usual gardening hierarchy or leap off the stifling naval career ladder, which depended so often on personal contacts. The job specification was somewhat daunting, and rarely advertised, and might have read roughly like this: 'Young man sought for botanical collecting. Knowledge of all forms of cultivation of tender species essential. Strong disposition a prerequisite. Must be willing to travel and of good character.' However, if those who won the jobs and survived to return home had written an advert it might have read something like this:

PLANT COLLECTOR NEEDED
for EXOTIC EXPEDITION
{apply on *personal recommendation* only}

CONDITIONS:
– Pay ABYSMAL, *discomfort* the norm.
– Sea travel essential, mostly courtesy of a WARSHIP (which may veer off to an *unstable* area unexpectedly and may possibly be away for an extra year).
– Assistance on the job *minimal*, unless you strike very lucky.
– Choice of TRAVELLING COMPANIONS limited, may sometimes help themselves to your goods, your finds or leave you *stranded*.
– LONELINESS *guaranteed*, mentoring non-existent.

- Answerable to BOSSES who might never have visited a *kitchen* or *laundry room* in their lives, and will certainly never have gone *grocery shopping*, but expect you to be totally self-sufficient.
- Essentially UNPAID, with a small *allowance* sometimes available for ESSENTIAL FOODS at times when it cannot be shot, dug up or fished for.

ESSENTIAL SKILLS AND PERSONAL QUALITIES:
- Excellent *knowledge* of BOTANY.
- Marksmanship with GUN of *highest* order.
- Technical *know-how* (repair manuals non-existent).
- Hunting, butchering, plucking and cooking.
- DRAWING and PAINTING of *highest* order (and ability to carry out same in *extreme* weather conditions).
- Long-distance WALKING (average rate of 4,000 miles a year).
- MOUNTAIN CLIMBING, MAPPING, COMPASS SKILLS, SNOWSHOEING, CANOEING, RIDING, OXEN-, HORSE- *and* DONKEY-DRIVING (tree climbing ability also *strongly* recommended).
- *Ability* not to be *mistaken* for a SPY (if you are, do not blame bosses, they will know *nothing* of your TROUBLES, being far distant).
- *Ability* to get on with PEOPLE from other ETHNICITIES essential, even if said peoples show *strong* RELUCTANCE to respond in FRIENDLY FASHION. (*Rapid thinking* necessary as said people might make clear a FORCEFUL URGE to dispose of you in SWIFT or *lingering* and MOST *unpleasant* terms.)
- Medical knowledge HIGHLY DESIRABLE (if survival of self preferred).

———

But of course such starkly truthful advertisements never appeared. The accurate tale of a plant collector's journey could only be written when, or if, he returned safely back home. Even then, what these returning explorers imparted was more often than not a sanitised account, approved in advance by their benefactors. These accounts concentrated on the botanical riches of these virgin lands and were told in breathless tones to the appreciative and admiring 'armchair adventurers' of the geographical societies. Word was then rapidly spread both by newspapers and by word of mouth, reaching the potting sheds of gardens up and down the land. The only negative note was sounded by the death toll: for every plant collector who retuned home with exotic flowering treasures, an equal number died on foreign shores. Nevertheless, the Scots kept on going.

Levering roots out of unforgiving soil the world over, these Scots found that their lives ricocheted from one bizarre or ridiculous situation to another. They might find themselves being route-marched to slavery in Afghanistan, disappearing for ever in the Californian gold rush, or being stood up by Empress Catherine the Great of Russia. They might get caught in the crossfire of vicious battles in the West Indies, shipwrecked in Cuba, or dumped in southern Spain as a prisoner of war. One endured witnessing his son being speared to death in remote Australia. Another was imprisoned by his captain on board ship. Others survived extremes of temperatures in the Arctic (with Franklin), in the Antarctic (with Ross before he went in search of the lost Franklin) and in broiling mid-Texan heatwaves. Exposure to the dreaded killer scurvy, or being forced to evade pirates in the East China Sea or operate a desperate shoot-to-kill policy against aggressive grizzly bears or marauding natives were also possibilities. However, it was always a reluctant finger that pulled the trigger, even in a life or death situation.

Tough as these men were, being at the same time the Indiana Jones of their day, and, in the main, lifelong wanderers, there were other sides to their characters. As they scrambled over virgin areas of the world, a terrible truth dawned for many and they foresaw with sorrow and clarity the destruction of these remote environments. They were never plant plunderers, ripping rare and precious specimens from the soil. Later, plant collectors such as orchid hunters would pluck the rare orchids for themselves, and have no hesitation in ripping up any surplus ones growing round about, so that any rival collector would be thwarted. Rather, they pre-empted the thoughts of poet Gerard Manley Hopkins who sang the importance of the wild places in his poem 'Inversnaid', inspired by a visit to Loch Lomond in 1883:

What would the world be, once bereft
Of wet and of wildness? Let them be left,
O let them be left, wildness and wet,
Long live the weeds and the wilderness yet.

Thus, these sensitive plant hunters shared many traits in common, such as self-sufficiency, intrepidness and plantsmanship, but why were there so many Scots amongst their number? There can never really be a true answer, but with certainty there are two factors common to all these Scots.

Firstly, the unwitting influence of one unlikely horticulturist and entrepreneur is important. Philip Miller, a London-based, cosseted, cocky and clever man, could well have been the spur that set this train of young men on the path to exploration. In the 1730s Miller was the doyen of the all-powerful Chelsea Physic Garden in London, a pinnacle of botanical knowledge and the point of entry for all new plants arriving in England. As the custodians of such a botanical powerhouse, and experts in cultivation of plants new and previously unknown, Miller's gardeners were in a position to acquire unique knowledge. During his custodianship of the Garden, the acquisition of plants from overseas changed tack. Up until this time many plants had arrived from abroad in the care of the naval physicians, whose searches on land close to where their ships docked would have been concentrated on finding likely medicinal herbs to replenish their dwindling chests. Now, instead of plants arriving by chance after being picked up from wherever a ship had happened to put in, organised searches were taking place with specialist young gardeners or surgeons being dispatched to specific corners of the globe to bring back useful and valuable plants. Miller found himself in the forefront of this change, as it was within his garden that a ready source of young men existed, trained in the ways of tender plants, and they were nearly all of Scottish origins. This was due to a curious quirk of his father, a first-generation Scot in London, who had insisted on employing only young Scotsmen, a habit that both his son and successors continued.

The second common denominator is, of course, that all these plant hunters spent their childhood and early adulthood in Scotland. They did not spring up from one particular geographical area. In contrast they emerged from seemingly unconnected backgrounds: they were brought up in areas ranging from Inverness to Old Meldrum in Aberdeenshire; Crieff in Perthshire to Auchinblae in Kincardineshire; Aberdeen to Glasgow, Forfar, and the Borders. In other words, they came from all over Scotland. The significant influence seems to have been national rather than local or regional. At no time was there a clutch of potential plantsmen produced, for example, from one island alone, in the way that Orkney produced so many employees for the North West and Hudson's Bay Companies in Canada nor from a certain region, like the Aberdeen men who served within a regiment or their local militia, the Gordon Highlanders, or the Perthshire lads who served within the Black Watch. The significance of the Scottish context seems rather that the Scottish gardening fraternity had its very own excellent 'bush telegraph'. With no gardening newspapers or magazines, which took some time to reach popular audiences, they nevertheless kept in touch with each other for news and developments. Word spread from the successful Scot who was head gardener for the great dukes and earls of England that he was looking for young men, and although one cannot discount the power of nepotism in the gardening world, it seems to have been the quality of the young trained Scottish gardeners that won the day.

So why were these Scottish gardeners so well qualified for the posts? Although Scotland appeared less sophisticated than England in the seventeenth and eighteenth centuries, it would be wrong to think that Scotland was a wasteland in terms of ornamental and productive gardens. On the contrary, there was a long history of wonderful gardens in Scotland. The earliest monasteries were founded in the twelfth century by monks who had come from France. These monks had brought with them much expertise about skilful fruit gardening, unknown in Scotland at that time. For many years this gardening knowledge was restricted to the monastic lands; owing to the instability of much of Scotland, gardening was a luxury rarely practised. Following this, by the end of the seventeenth century with the onset of more peaceful times, Scottish aristocracy built houses which had less of a requirement for fortress-like constructions. Gardens sprang up around many great houses, clustered around Edinburgh and what is now East Lothian, spreading through Fife, skimming up the coastlines of Angus, Kincardineshire and Aberdeenshire and on to the Moray coast. Their distribution mirrored that of the productive farmlands that had produced the riches of the owners in the first place.

These new gardens needed gardeners and there is plenty of evidence that it was a fairly attractive position. The way in which gardeners were employed was redolent of a sub-contractual agreement with precise clauses. Generally the head gardener was allocated a house with field, providing grass enough for a cow or two. Sometimes winter fodder was thrown into this equation and oatmeal amounting to several bolls, one of which weighed 140 pounds (64 kg). In 1653 gardeners appeared to be paid about 33 to 40 Scottish pounds. By the 1720s they were paid almost double that.

The agreements between gardeners and their employees ranged in detail from very exact documents in which under gardeners could find exactly how much fruit and vegetables they could take out of the garden for their own use, as well as perhaps the thinnings of the woodland for firewood. Sometimes they were allowed to fish. Generally a head gardener earned double the amount of his assistants and the apprentices were only paid token amounts because they were there in order to learn their trade. In the hierarchical pecking order of the staff for big estates and mansion houses, gardeners were often paid roughly the same as the head person in the household: the cook, housekeeper or butler.

However, a head gardener was required to do an enormous amount for his money and this commitment was often felt to be undervalued. The fashionable status symbol for the man of substance at the time was a French cook, upon whom many lavished generous amounts. The respected Edinburgh nurseryman William Butcher raged that 'a great man bestows from 50 to 100 pounds on a French cook, but to a British gardener seldom more than 20 to 40 pounds'. But the fashion for French cooks ebbed whilst that of the gardeners was consolidated.

The rising and steady demand for young Scottish gardeners may have also been due to the consistently high level of training they enjoyed, a training that owed much to the high standards and commitment of these head gardeners. A head gardener's tasks did not end at the close of the day. In 1813, it was noted that the education given to these young apprentices by their superiors included the essence of the 'three Rs' – reading, writing and arithmetic – as well as botany, willow basketry and even making mole traps. No wonder then that this grounding set many young gardeners on a higher level when it came to applying for jobs in England. Scottish education at elementary level was more widely available than in England, and so the youngsters could more readily absorb detailed knowledge about botany, which required reading skills.

Thus, it was hardly surprising that the Scots who already held powerful head gardener positions in the south of England chose their fellow Scots as assistants: they could rest assured that the youngsters would arrive with high standards in literacy and plantsmanship. Indeed Scottish training and the Scottish bush telegraph were so successful that several gardeners went down south only to be enticed back by the great Scottish lairds, who were so desperate to employ a good Scottish gardener that the gardener himself could sometimes set his terms. The dukes and earls would cast around amongst Scottish seedsmen and nurserymen in England for likely recruits and these nurserymen in turn acted as unofficial recruitment agencies.

Despite counter-current, in the early part of the eighteenth century the flow of Scottish gardeners down south was so great and they were so well respected that it got to the point that employers would assume that any head gardener at any major garden was a Scotsman. As the novelist George Eliot commented, 'a gardener [was] Scottish as a French teacher [was] Parisien'. It was even rumoured that if two men applied for a job in England and one was Scottish he would automatically be offered the post, and when he was established he would almost certainly teach his sons all he knew, thereby establishing a mini-dynasty of Scottish gardeners. As many of the nurserymen were Scots as well, they tended to trade with each other, and information would pass by word of mouth regarding good or bad employers, and gardeners ready to move on to better jobs.

Aside from the esteem in which the profession was held, the growing opportunities within Scotland, the good apprenticing system and the effective bush telegraph, what other advantages might a Scottish background have given gardeners? Many Scottish backgrounds included a type of crofting experience, which ensured that the young men were able to turn their hands to a wide range of skills, from growing crops and herbs, to ditching, wall-building, thatching, animal husbandry, and other saleable skills, such as weaving, cobbling shoes or butchery. Crofting produced the type of self-reliant character who possessed problem-solving abilities, and great resilience. So a large proportion of the gardening Scots were well used to managing a smallholding, the equivalent of running a small business at the same time as holding down a full-time job. As the working conditions south of the Border were often negotiated in post, Scottish gardeners flourished; they were often given the right to sell surplus fruit and vegetables, and supply plants at the local markets, and they possessed the perfect background to exploit these opportunities to the full.

Alongside these benefits, an upbringing in a rural and crofting background gave many Scottish gardeners a sense of self-sufficiency. When you lived on a croft, which would only give you at best a part-time income, you knew you had to turn your hand to other tasks. You absorbed skills, such as how best to till the land and produce a crop, be your own vet, mill your corn and butcher your animals, grow your own medicinal herbs and learn how to use them, tramp as a journeyman to another area and practise or learn another trade, frequently speaking English as well as Gaelic. From a young age, you were as familiar with the art of bartering as a black marketeer. You had learned to turn your hand to many tasks, your tongue to several languages, and you knew, if push came to shove, you could survive by quickly absorbing a new skill. Thus, Scottish gardeners were better educated from early childhood than most of their contemporaries across Europe in how to merge into another country, and profit from new opportunities.

From such backgrounds came these intrepid, dogged and quick-witted Scots, who brought back a legacy that, over time, has given huge pleasure to millions. It has also made more than a handful of millionaires.

Despite their common Scottish background, the plant collectors themselves were a varied bunch, and their exploits were rough and ready, and certainly opportunist rather than meticulously organised, rather like the characters themselves. Not once do we find a plant explorer who was dispatched on a journey as carefully mapped out as an early Thomas Cook holiday. In tracing their various turbulent voyages of exploration, very little of any cohesion stands out clearly. It would have been much neater to be able to parcel them up with a clear listing of where they went, why, and sent by whom, rather like bundles of clearly labelled seeds. Then one could open the neatly sealed-up packages, and have the map of their endeavours and results clearly visible. If only one could slot them all together like an easily digestible history book, listing generations of kings and queens with the outstanding dates, events and achievements (or not) of their lifetimes. One cannot.

But with the comfortable passage of time, now at least a century or so, their adventures do settle down into stories with a beginning, middle and end, and those extraordinary tales make up this book. Without doubt they brought back hundreds upon hundreds of plants that changed our gardens and commercial forestry enterprises for ever. They changed our landscape as surely as industrial giants changed our built landscape, but mercifully with a kinder impact.

The methods by which many of these plants reached these shores, though, remain colossally mixed and muddled. Some collectors were in the pay of rich or aristocratic benefactors, some of whom indulged in a passion of plants, and oneupmanship. Others were in the very grudging pay of the committee supremos of horticultural societies or the great botanic gardens of the nation. Others merely happened to gather as a hobby, were posted to far corners of the globe, and were only too happy to point a visiting collector to an area ripe with plant promise. Many more gathered plants as an aside from practising medicine in remote parts of the world, usually as a paid, serving member of the Royal Navy, where duties ashore would include finding plants which would replenish the herbal medicine chests, and would add to the knowledge of the Chelsea Physic Garden. Others were self-financed, usually pitifully rewarded, and constituted little more than the flotsam at the lesser end of the plant hunting hierarchy. But if the plant collectors' reasons for being collectors were rarely clear-cut, neither is the classification of their finds.

There will always be a touch of the lucky dip in verifying who should get the medal for being first past the post. It is all too frequently hazy as to who exactly brought back what and when. Who managed to land back on our shore a plant which was the same species but stronger or easier to propagate? Who got there first? Did someone send his first plant to the benefactor who had sent him out in the first place, but a second to his 'home garden of origin' so to speak? If the first benefactor simply paid no attention for a year or so, but the second produced a stunning plant, who actually had the first prize? Especially in an age when communication was erratic and protracted, these matters are not easily verified.

From as early as the seventeenth century, dried specimen plants began arriving home. Wherever they landed, most plants finally reached the British examining rooms of Kew and the Chelsea Physic Garden. Sometimes no one knew quite what these plants were, but once they were identified that constituted a 'sure find' and the lucky fellow jumped the queue and was elevated to being the first to find a plant, though he couldn't necessarily be credited with bringing it back alive. Being 'credited with discovering', with its overtones of academic niceties, was all too often distanced from the next step, successfully dispatching a handful of live plants or viable seeds back to base.

Success in bringing back live seeds was sporadic. One plant collector might send back a bare minimum of seeds, and only some might be coaxed to life and carefully passed on to the great aristocratic or botanic gardens of Britain, to be cherished as valuable curiosities, and confined to the privacy of these gardens, where they were jealously guarded. One of the main remits following the establishment in 1804 of the Horticultural Society in London (it became the Royal Horticultural Society in 1861) was to raise funds to sponsor their own collectors and send them out into likely parts of the world. Many of these plant collectors, who managed to send back armfuls of seeds and therefore facilitate an explosion of plants on the market, might well claim to be establishing 'a first', as it was the first time they had really been available to all and sundry. Certainly the nurserymen thought so. Some sent back seeds which those at home had little idea how to grow, in what conditions, and what the end product might look like. After all, a healthy semi-hardy plant like a *Pelargonium* (mistakenly but commonly called *geranium*) growing effusively and waist-high in South Africa, might look a shrivelled, slow-growing specimen in Scottish summers. It was ever a chancy occupation. Sometimes all seeds failed. Often the boxes failed to arrive. Many were lost at sea, or ruined by salt water. Sometimes the valuable cargo of plants, gathered with almost unimaginable effort, simply lay abandoned and gathering dust in the sheds of the great botanic gardens. No one had the time or the knowledge to deal with them. Sad tales abound of collectors surviving harsh months of searching, living rough and cheating death at the hands of inhospitable natives, disease or starvation, only to arrive home and find the box they had sent two years before lying untouched.

Yet in all these haphazard sagas, there were many times when the surprise and pleasure at the flowering of a wondrous, brilliant brand-new plant generated such excitement that cash was sucked in for another expedition, and off went another eager hunter to satisfy the lust for more and more plant wonders.

First, cash had to be secured. The Horticultural Society was one of the first to finance journeys by their own plant collectors to their chosen specific area of the globe. Before this, plant collecting had depended on, for example, British government officials stationed on foreign shores with a keen and abiding interest in botany. They would keep in correspondence with men of like mind, such as Sir Joseph Banks, a naturalist and scientist who had already circumnavigated the world with Captain Cook, or the gardeners in charge at Kew or the Chelsea Physic Garden. Alternatively, plants came back courtesy of naval physicians, whose training had included botany, and in which they retained a strong interest. They would scour the areas upon which they landed for plants and again correspond with the likely gardeners at Kew or Chelsea. All this was a haphazard and erratic method of gathering plants, but now ships were proceeding more rapidly around the world due to advances in design and sail power, and allowing previously unknown areas to be explored and exciting finds to be reported. However, this was only as and when the ships happened to be going there for other purposes. For the most part, only the king had been able to afford to mount his own collecting missions so far, such as sending Francis Masson to South Africa upon the recommendation of Sir Joseph Banks. What the brand new Horticultural Society wanted to do was to send its own collectors precisely where they would be able to gather plants, both for themselves, but also for commercial gain. They wanted plants to propagate and sell. When the Society's blood was up, this train of events from initial idea to action was rapid. For only a few months to elapse from calling a committee to waving off a young man was not unusual. The committee of the Horticultural Society of London would simply pinpoint a vessel making for the correct destination and secure a willing recruit. And of the latter there was an inexhaustible Scottish supply.

Claims and counter-claims of who brought back what and when are legion. Jockeying for position behind the scenes was understandable, but when success dawned it was subject to the very best of public relations propaganda and – possibly – exploitation. The hardships and brushes with death involved in an average expedition transformed a couple of returning plant collectors into newsworthy public property. These unlikely celebrities were catapulted blinking and dazzled from their solitary harsh lives into the splendour of the great rooms hired by the geographical and horticultural societies. They enjoyed brief weeks of being fêted. Their fame certainly reflected well on their benefactors, and these benefactors were masters of spin. Such star treatment was enjoyed – or endured – by some returning collectors, but the reward for many was far from immediate glory and fame.

Poor, young John Jeffrey from Perthshire, who rose from farmhand to explorer within a few years, was dismissed by his supporters, the 'Oregon Association', who comprised the good burghers of Edinburgh. They had wheedled sponsorship from Queen Victoria's Consort, Prince Albert, as well as dukes, marquises, earls, knights and prime ministers. Jeffrey was deemed to have let them down by his lack of acceptable collecting prowess when trawling his allotted patch of the west coast of the USA. Aged just twenty-four, he had been required to collect seeds of the great conifers of the area, as well as suitable herbaceous plants which caught his fancy, and to keep his eyes peeled for unusual beetles. Before he could achieve this, however, he had first of all to get over a substantial tract of Canada on foot and canoe with the tough men of the Hudson's Bay Company; this entailed a journey of several thousand miles and then a climb through the Rockies to the Oregon area.

This he achieved, and he was then responsible for sending back amongst others the exquisite *Dodecatheon jeffreyii*, or 'Shooting Star' type of dog's tooth violet, *Iris tenax*, a much-used iris today, as well as lilies, lupins, and penstemons. Growing in the grounds of Highgrove in Gloucestershire, the home of the Prince of Wales, is a wide-sweeping ribbon of purple tulips interplanted with a Jeffrey discovery, *Camassia leichtlinii*, a beautiful, tall, deep-blue bluebell look-alike. All in all Jeffrey collected 199 species. But the committee men of the Oregon Association, spread round the civilised drawing rooms of Edinburgh, were more than displeased with the lack of letters home, and the large gaps between boxes arriving on British shores. In their discussions, they sound like exasperated parents trying to make sense of an uncommunicative, recalcitrant backpacker. Why wouldn't he respond? Reluctantly, because there was a *frisson* of worry about his safety, and perhaps because they were embarrassed to admit that they might just have chosen the wrong man, they decided to terminate his contract. Jeffrey never received the news. By twenty-eight, he was both dismissed from a distance and dead.

Communication with home was drawn out. Letters home, if they reached there at all, might take a year. Ships around the world would be greeted frequently by figures racing down to the local port, waving an epistle sealed with a great glob of wax, anxious to find out the destination of the ship. The bush telegraph was rapid. Time and time again, one of these Scottish plant collectors would abandon a trip up-country on the whiff of a rumour that a ship had arrived. Nor were they simply longing to hear news from home: a frenzy of packing would be undertaken of seeds and plant trophies, the dried skins of curious beasts, fruits pickled in spirits, fossils, and fir cones along with scratched letters of explanation.

Around the world there were enthusiasts. British diplomatic envoys and commercial representatives scattered round the globe by the eighteenth century onwards often took a great interest in the natural world around them and established correspondence with like-minded men at the great botanic gardens of Britain, at Kew, the Chelsea Physic Garden, the Botanic Gardens of Edinburgh and Glasgow, to name but a few. They accepted the tardiness and haphazard nature of letters written and answered. With round robins of letters and information, these plant-crazy pen-pals kept up lengthy correspondences. They mirrored the many clergymen who were stuck in rural backwaters with too little to occupy themselves, few of intellectual equal with whom to converse and a lifetime of servitude ahead. So they sought refuge by absorbing themselves in the ever-changing natural world around them, and sometimes indulged in crafting great illustrated tomes on plants and their habits. It was an occupation which appeared a suitable, genteel and honourable pastime, of unimpeachable moral principles.

These varied correspondents cut across continents, oceans and class. Plants had no social airs, responded not a whit better to the hands of a baronet than a peasant. In the mainly rural lives of most British people of the eighteenth and early nineteenth centuries, just about everybody had a patch of earth available to them. In the first decade of the nineteenth century, an avuncular Bishop of Carlisle kept up a three-way correspondence with the very eminent Dr James Edward Smith, founder and then president of the Linnean Society, and George Don, irascible Forfar gardener and son of a shoemaker. And it was Don who rasped out his displeasure at lacklustre responses, and had no hesitation in taking issue with what he detected to be an incorrect identification of a new specimen.

In many corners of the globe, where ships bearing many of the plant collectors called to refuel with trading goods, water and fresh food, there was an astonishing number of these correspondents and plant aficionados. Many nurtured local plants and were only too happy to give generously of their seeds. Glory was never sought, generosity of spirit was greatly in evidence. Perhaps it was enough that it was an escape from local politics, unstable countries, and the ever-present fear of uprising and disease. In the 1950s, Vita Sackville-West, the creator of one of the great gardens of England, at Sissinghurst, referred to the giving spirit of 'that great freemasonry' of gardeners. Seeds or cuttings would be given gladly, taken on board, but only grudgingly offered a small space. After all, of what importance were a few indeterminate plants thirsting for daylight and protection from salt water, compared with surviving scurvy and sailing round Cape Horn?

Against so many odds, some seeds survived. Those who unwrap seed packets mistakenly left in a damp potting shed over the winter might empathise here. Is it easier and more certain to throw out a mouldering packet and buy afresh? But if those seeds or plants were likely to be the great new 'discovery' for the nurserymen back home, endless care might be taken. Seeds were wrapped up in paper, which then often allowed damp to seep in. Out in the field, the damp paper would be peeled off, and the seeds dried off and repacked. This cycle would be repeated over and over again in order to preserve them for as much as a year at a time. Despite these adverse circumstances, a generous proportion of the seeds and plants that were lugged back to Britain took on a life of fame and fortune of their own.

By coincidence, one who produced through his plant discoveries a ready source of wealth around the world, actually bore the surname Fortune. Robert Fortune's life, from his birth in Berwickshire on the Border, in 1813, to his death in 1880, encompassed some fame, a very small fortune for himself, and a swashbuckling life of adventure that he certainly did not seek out. Sponsored and financed by the Horticultural Society and then the East India Company to search for new plants, he dodged death in a multitude of imaginative ways. He disguised himself as a Chinese man in the days of the mid nineteenth century when China rewarded most foreign visitors with torture and death. He created the ultimate naval ruse, by forcing his trembling Chinese boat crew to dress as Europeans when stalked by pirates, thus discouraging his would-be assailants who regarded armed Europeans as certainly more to be feared as marksmen than their brother Chinese. And despite this blood-splattered, hazardous life, he managed to send back to Europe some of the most fragile, exotic and desirable flowering plants. Many of his finds, such as camellias, are a commonplace part of modern life. Robert Fortune also contributed to the chrysanthemum-growing craze by sending back the rather less showy but appealing pompom versions. He would certainly have approved of the National Chrysanthemum Society, established in Stoke Newington in 1846, when Fortune was thirty-three.

Fortune's story reinforces the impression that a steady stream of organised expeditions was not the norm. It was a haphazard collection of Scots – likely lads at best, accidental misfits at worst – who departed under a deluge of instructions from one of the various organisations, usually controlled by comfortable desk-bound male committees who saw a fast buck in the offing and drove the hardest of bargains in order to secure a financial return. If these young plant explorers were prepared to work more or less single-mindedly around the clock for at least a year or so, their reward, should their life, if not their health, still be intact, was not fame and fortune, but rather a paltry sum in the bank.

Survival did not even honour David Douglas, the son of a stonemason from Scone, whose life ended when, aged thirty-four, he stumbled into a trap for wild cattle on one of the Hawaiian islands. In ten years he had collected almost more plants than any other single explorer, and his legacy still grows. The majestic firs, with the tongue-twisting Latin name of *Pseudotsuga menziesii*, are known throughout the forestry world as Douglas firs. Other finds which have produced a bonanza for growers are Sitka spruce. Attractive trees, Douglas mused when he found them. How disappointed he might well have been to view them on Scottish hills, packed into sombre rafts of rectangles. If he was like so many other young plant hunters, his idealism would not have evaporated; for most their respect for nature grew. Indeed, half a century before John Muir left Dunbar for California, and triggered a challenge to the despoilers of our planet, Douglas was questioning the destructive path of the fur-trapping Hudson's Bay Company with their policy of total destruction towards the beaver. Hundreds and then thousands of miles were left bereft of these animals, prized merely for their pelts, and even that only while the fashion lasted.

Almost all set off with a sense of destiny, honoured in being chosen, strongly determined to succeed. Few explorers set out as married men. None expected to meet their prospective wives in such remote areas of the world when out collecting. Rather than making their fortune, most simply hoped to return with even modest success, thus enhancing their career chances. Some could not settle down, others married late in life. But if little is known regarding the women in their own lives, we do at least know that women embraced the new varieties of plants they were bringing back.

Leisure time meant that comfortable middle-class ladies espied curious pictures coming back from the Far East, and wanted more of the extraordinary plants depicted on incoming pottery, fans and textiles. Colour had always played a huge part in decoration when affordable. Tropical colours rained down on the nation's gardens. Rainbow colour schemes, of ever more vibrant or curdling colour clashes, depending on your point of view, gave a brand-new look to the bedding schemes of parks and gardens. Crowded together so that the soil vanished from view were brilliant scarlet *Pelargonium*, brought back by Aberdonian Francis Masson, who endured snakes as sleeping companions when roughing it in South Africa. *Aster novae angliae*, Michaelmas daisies, and varieties of *Coreopsis* sent over from North America by Thomas Drummond added to the scene.

From the mid 1750s to the end of the nineteenth century at least, the world of British exploration for plants was one of our great success stories with collectors dotted all over the world, although rarely numbering more than a handful at any one time. Dumped off in largely inhospitable parts of the globe, in climates they had never experienced, searching for plants they had rarely seen, at best relying on their wits, elementary medical know-how, and mostly minus any map, the diaries which they faithfully kept evolved into embryonic 'rough guides'. Keeping a diary was a condition of the contract.

Many diaries still exist, locked safely away within the great establishment archives and botany libraries. Each is around A4 size, with firm cardboard covers embellished with marbled paper. The writing is at best spidery and crowded, as paper was at a premium. Many a time the writing turns sharp right angles, like a Roman road, and continues up the margins, making deciphering difficult. The paper itself is often tissue-paper quality. Guesses with question marks hover many a time over puzzling plant finds. But it was one of the conditions of the trip, keeping up that diary and just like schoolboys with tiresome homework, they kept them up, possibly out of nervousness at being found out if they didn't. Just like schoolboys they often noted just how tedious all the writing was, so much better to be out and about exploring.

Sometimes, as I turned the feather-light, fluttering pages I pondered on what would have been the young men's thoughts as they searched for words and expressions. Written at sea, on laps under upturned boats, beside bonfires in forests with ever-present danger lurking and exposed to the elements, there are surprisingly few words scored out. Pen and ink did not have a delete button. Most clearly tried hard to get it right first time on the page. Or was it rigorous Scottish educations that had driven in such discipline and accuracy?

In place of 'where to eat' they noted what the locals ate, and tried all manner of foods, sometimes with serious objections from their digestive tracts. In place of 'what attractions to see' they climbed the odd mountain, or slashed their way through jungles to reveal stunning views or be stunned by eyeball-to-eyeball encounters with aggressive animal or human predators. In place of 'languages required' they developed a fast line in mimicry, as that longed-for scoop, the seeds of the most desirable new acquisition for the great royal botanic gardens of Britain, might be in close proximity, but there was little time to hang around as their boat would be sailing soon, and another might not arrive for months, or even years.

It is astounding that so many diaries survived. It is also astounding just how much in terms of plant life was brought back. In the great scrabble for valuable seeds to satisfy a voracious and growing appetite for plants, Scotland produced successes out of all proportion to the nation's size. The numbers do match a Scots rugby team, and like so many glorious teams of the past, few except for the diehard enthusiasts can remember more than a couple of names.

Alongside these bigger names, many more Scots made at least a noteworthy contribution to gathering the plants we take for granted in our gardens. This book elaborates the lives, expeditions and legacies of the main explorers, but there were many, many more besides, for whom sadly almost no substantial records remain.

CHAPTER 1

Philip Miller (1691–1771)

THE SCOTTISH 'GODFATHER' IN LONDON

———

Philip Miller developed from an apprenticeship with his Scottish father, a successful entrepreneurial nurseryman, who had arrived in London from Scotland in the late 1600s, into an influential and dogmatic head gardener for the Worshipful Company of Apothecaries, whose garden, the Chelsea Physic Garden of London, he controlled and ran for decades. He was also one of the most knowledgeable gardeners of his day: not only could he recognise a wealth of plants, he kept up a worldwide correspondence with other gardening enthusiasts. He kowtowed to no one.

He was innovative, experimental, bold and a trend-setter. It took great self-confidence to try out solutions which others might fight shy of, and push forward his futuristic ideas. And it paid huge dividends. He described how to force bulbs in vases of water, like the 'hyacinth glasses' still popular today; he coaxed into maturity a tall, lanky, American tree, *Abies balsamea*, from which is still extracted an essence used to stabilise perfumes. When forced to transplant some damask roses just as they were about to flower, he chopped off all the flower buds, stuck the roses in well-watered trenches and was able to pick their blooms later in the season.

He was a guiding and dominating force during an explosion

of interest in all things botanical in the mid eighteenth century. At the receiving end of a trickle of seeds from exotic locations around the world, he guarded their propagation and was determined to be one step ahead. He was the gardening guru of his day and a couple of hundred years later, he would have been a television superhero. *The Gardeners Dictionary* opens with a frontispiece which announces that the book is written according to the practice of 'The most experienced gardeners of present age', by which he means himself. By the time of his death in 1771, not only had his massive *Gardeners Dictionary*, first published in 1731, been reprinted eight times, even Linnaeus, the Swedish botanist responsible for classifying plants into family groups, had conceded that it was a botanical masterpiece as well as a 'know-all' about the craft of gardening.

But how had Miller come to be in this position? Philip Miller was a second generation Scot and a much cherished son of a successful nurseryman. His father had trudged down from his native land and worked as a gardener at Bromley before emerging a few years later as a successful businessman, having a thriving market garden. His father had arrived in London well before Philip's birth in 1691, and his endeavours to better himself were immediate, and fruitful. In the time-honoured way of the self-made man, he cut no corners in ensuring that his son received the best, and Philip was very well educated. Not for his son a cursory schooling until the age of twelve, then apprenticeship, with the prospect of a lifelong slog in a garden. Young Philip was accorded the hothouse treatment by an indulgent and determined father, with little choice of career.

Pehr Kalm, a Swedish-born professor of botany and a great naturalist, visited England in 1746 on his way to North America, and recorded much about his English visit. His diaries include a summary of Philip Miller's education.

He began to instruct his son in the art of being a nursery man from his earliest years, as he had an uncommon liking for that occupation. As the man throve, so he spared no expense in also causing his son to have a sufficient education in various languages, and other sciences, which profit and adorn a man. Miller quickly assimilated all that his father had himself taught him, within that theory and practice of ornamental and kitchen gardening. At the same time he went through all books which had appeared in England on these sciences. An industrious intercourse with other enterprising nurserymen in this town and in the country round about made him still more proficient. But he did not stop with this.

His thoughts were therefore turned upon travelling. He was well off and had therefore no difficulty in accomplishing this. To travel out to foreign countries without having first made himself acquainted with what remarkable things there are to be found at home, he held neither for wisdom or usefulness. He therefore travelled through the greater part of England, observing everything, but was equally careful to inspect all ornamental and kitchen gardens, and to make himself at home and acquainted with all horticulturists, for he was of the opinion that he could learn something useful which he did not know before, from at least some of the gardeners he visited . . . As agriculture had so near a connection with horticulture therefore he kept at the same time an observant eye on everything which occurred in rural economy, particularly the cultivation of ploughed lands.

After he had travelled through England he started on his foreign travels and thus explored Flanders and Holland, because he knew there were also great horticulturists there and with the science and management of ornamental and kitchen gardens which there reached a high pitch of excellence. Where he, besides the aforesaid lands, also explored other districts, I have not understood.

Philip Miller proved an admirable spy in horticultural espionage. He had gained a tantalising glimpse of the cutting edge of horticulture in the country which spawned tulip mania, and saw that glory and fortune were there to be made.

Having also passed a leisurely few years deciding on a future career, travelling around both England and the Low Countries, a luxury afforded to him by his well-to-do father, Miller made up his mind. He had not ventured north to pay Scotland a visit.

The news in London would have made a tour of Scotland appear very unattractive indeed, if not dangerous. Stories in the press from north of the Border were sporadic and, in the best tradition of journalism, weighted to the more interesting news, which would have been generously sprinkled with dire happenings and disaster. The devastating famine of 1709, just after the Union of the Crowns, had resulted in poverty for much of Scotland. Miller's journeying north would have coincided with the 1715 Jacobite Rebellion, and the unrest would have been well reported. English news would have painted Scotland as a backward, disorganised, and dangerous country, and one best not to venture forth in. It was also a seriously inauspicious time for many Scots to venture at random into England. Images of Scots on the make, or whingeing Scots were popularised in cartoons.

So having ruminated on the horticultural career possibilities open to him, Miller plumped for cultivating flowers for ornamental gardening. It was a wise decision. Having toured Holland, he had noted well the art of forcing for out of season fruits and flowers, a skill brought over with the arrival of the court of Dutch-born King William in 1689, but not yet widely practised in England, requiring as it did copious glass, which was an expensive commodity, and accurate temperature control. The latter was a highly skilled task, and the Dutch still very much had the edge in necessary expertise. It was a skill he had been able to observe at first hand. Miller was a highly intelligent and inquisitive man, and he saw an expanding niche in the market. He decided he could fill it rather well. With his father's generous pocket, he opened up his own nursery at St George's Fields, Southwark. However, although he basked within familial financial comfort, he had absorbed a strong work ethic from his father. His nursery flourished, and his name, in his own right, attracted attention. He appeared to be safely on an upwardly mobile path.

John Rodgers wrote in his book, *The Vegetable Cultivator*, in 1839 that Miller was the son of a market gardener and, at the time of his appointment to the Chelsea Physic Garden in 1722, was the owner of a flourishing florist and ornamental shrub nursery in London. He was recommended to Sir Hans Sloane, who picked him as the ideal candidate to become the Chelsea Physic Garden's head gardener. This was indeed both a wonderful stroke of luck, and a just reward for all his hard work. It was also good timing, as the lease on his premises was coming to an end. Sir Hans Sloane had very much picked the right man, at the right time, for the right job.

Not only did Miller's wise and confident hand guide the gardens to become one of the most famous and prolific botanic gardens in the country, he corresponded with European plantsmen and collectors in America. As part of the web of collectors he also played a key role in the great horticultural expansion which was beginning to enmesh the globe.

Miller was not only a botanist of formidable knowledge, he also nurtured plants from all over the world. They arrived, often more dead than alive, frequently in a fragile and battered state. As *The Apothecaries' Garden* summarises: 'As for the ideal growing conditions, he could only surmise, as many of the countries from where they came would have been as alien as another planet. He possessed a genius for finding solutions. He grew pawpaw, melons and pineapples in beds of fermenting oak bark discarded from tanneries, and his fame spread, even to the table of the king who graciously accepted a pineapple presented by Sloane.'

In order to remain dominant, however, Miller needed a succession of willing, trained subordinates. Like millions around the world who cling tenaciously to their Scottish origins, he never set foot on Scottish soil, so it must have been a purely inherited Scottish thrawn (obdurate) streak which led him to insist on employing only Scots in 'his' gardens. One of the reasons such a ready supply was available was the existence of not just one, but three physic gardens in Edinburgh: at Holyrood; in town at Trinity House; and at the university. Before Miller was a twinkle in his father's eye, way back in 1683, James Sutherland from Edinburgh, a head gardener at the Physic Garden in Edinburgh, had published a much revered *Hortus Medicus Edinburgensi.*

In fact, Scot after unremembered Scot had already made his mark in the steadily advancing gardening explosion. By the late sixteenth and early seventeenth centuries, many of the great Scottish gardens, which had often been created well before the Reformation in 1567, had vanished, or become starved of cash, owing to the struggling Scottish economy. Only amongst the hugely wealthy who benefited from the 1707 Union of the Crowns was any form of ambitious garden scheme under way north of the Border. But Scottish gardens had once flourished and had provided a training ground for a multitude of gardeners. So, in the early seventeenth century, after the Union, there was an army of Scottish gardeners ready to march south. In disdain, many English complained about this invasion. But the Scots were well trained, had often been 'finished off' by working at the great botanic gardens in Edinburgh or Glasgow, and, owing to the network between these gardens and the Chelsea Physic Garden, they simply walked into the best jobs. One can understand only too clearly the strident cries of protest from the English.

The opportunities afforded by this web of Scots in charge of the greatest gardens in the land, in London, and the great gardens of the hugely wealthy aristocrats in England, acted as a honey pot. Just like cooking with out of season or 'new' vegetables and fruit from foreign lands, who would not have wanted to get their hands on the cultivation of 'new' and exotic plants from all over the world, growing them under the great glasshouses which only the very wealthiest in the land could own? And these treasures existed almost exclusively in England. It's a tale whose shape is still familiar today.

Miller, meanwhile, was still advancing his own career. He had a unique knowledge of the successful cultivation of 'exotic' species because of his position at Chelsea Physic Garden and he was not slow to realise this was a lucrative skill. He took with ease to writing, and his *Gardeners Dictionary*, which had attracted 400 subscribers on its first publication in 1731, covered every topic of gardening, from fruit and vegetables, to vineyards and greenhouses.

It was translated into Dutch, German and French. He was the horticultural Mrs Beeton of his day and his books, praised for their wide breadth, were a publisher's dream, often running into many editions. With a calculating eye for markets, he also produced a budget abridged edition for the middle classes.

But all was not running so prolifically in the garden, which was being managed in a lackadaisical fashion by the Society of Apothecaries, to whom Miller was answerable. The minute books resided variously in the Garden offices, at the nearby Swan Tavern, and eventually at the Society's hall in Blackfriars. A simmering, increasingly bad-tempered quarrel culminated in Miller's dismissal, when aged seventy-eight. One of the criticisms levelled was that in essence he took upon himself the role of manager and director. In the light of the bad management noted above, it is easy to see why he might have done so.

A success not only within the garden, but as a published author, Miller saw himself at the cutting edge of botany and gardening. The well-educated young Turk of earlier years, with his wide interests, great skills and encyclopaedic knowledge, had probably grown too big for the boots the Society required filling. The goodwill between Miller and his employers had been sullied by petty disagreements, such as disputed ownership of some orange trees. They preferred someone a little more malleable, a little less well educated, less opinionated, and in short, more of a straightforward, well-trained gardener. It appears that they preferred to make the more cerebral decisions themselves, while their employee kept himself confined to looking after the garden. So, Philip Miller's long and illustrious reign at Chelsea came to an end in 1770 when he was seventy-nine years old. Perhaps equally galling, a young apprentice, whom he had hand-picked, became his successor.

CHAPTER 2

William Forsyth (1737–1804) and William Aiton (1731–1793)

PUTTING SCOTS FIRST

William Forsyth had fulfilled the many exacting requirements of Miller. Most importantly, he was Scottish and he was a hard worker. Forsyth's early years, from his birth in Aberdeenshire in 1737, apparently at Old Meldrum, until his appearance in 1763 as a gardener at Syon House, from where he rose swiftly to the Apothecaries' Garden at Chelsea, are very much based on hearsay. Not for Forsyth the carefully choreographed upbringing within a well-to-do home. He had more than likely tramped the entire country, but not, like Miller, as a gentleman observer. He seems instead to have been a shrewd lad from an artisan home, with an eye on a better life. In 1763, as he covered mile after mile down from the north-east of Scotland, eventually settling down amid the sumptuousness of the grounds of Syon House on the banks of the Thames almost opposite Kew, he epitomised to the letter Dr Samuel Johnson's acidic remark of that same year, 'The noblest prospect which a Scotchman ever sees is the high road to England!'

Reputedly, although there is no direct evidence for this, he worked at Haddo House, the newly built mansion erected by William Gordon, the second Earl of Aberdeen, a 'thumping snob' according to his descendants. History records him as an exploitative landlord and an acquisitive neighbour. If Forsyth did cut his teeth gardening at Haddo, it is likely that he was apprenticed in the kitchen garden, which would have been required to grow copious quantities of fruit and vegetables. Again, unlike Miller, he had precious little choice when it came to nurturing fruit, vegetables or flowers. Food for the earl's table was to be coaxed out of the land or dismissal followed. If the weather was a constant enemy, at least the soil was rich, Haddo being set amid highly desirable agricultural land. In any event, the toil would have been lightened by news of the local goings on. Tales of ruthlessly acquired riches by the most powerful men in the area would certainly have been a major talking-point in the area, and with embellishments in the potting shed, they would have been ample to keep him going on the long tramp south.

What effect might these tales have had on young William Forsyth? All the evidence points to him possessing a deep-rooted and ever canny sense of survival amid the aristocracy. As he turned over the soil at Chelsea, pricked out curious seedlings from distant lands, and kept his ears open to instruction, reward was always on hand. At the age of thirty-four, on 6 February 1771, Forsyth's square form tramped over to the offices at the Chelsea Physic Garden to formally receive the keys. Who knows how he felt about profiting at the expense of his boss, Philip Miller, who had also been his tutor and mentor. Miller was dead within the year. Forsyth, though of lesser education, and certainly a less privileged background, had a steady head for power, prestige and money. He remained at the garden for thirteen years.

The contract offered by the Society was one created from lessons learned, and he was required to be much more accountable. The Society was wary of employing another gardener whom they perceived as a loose cannon. Forsyth presided over many improvements at the garden, and his diplomatic skills stood him in excellent stead. On the one hand he was receiving 500 plants from Sir Joseph Banks' round-the-world trip with Cook on the

Endeavour; on the other, he kept on good enough terms with his previous employer, the Duke of Northumberland at Syon House, and from a Dr Clerk in Jamaica he took possession of the allspice tree, *Myrtus pimento*.

Such was the challenge of keeping innumerable plants requiring tender care, that he was soon complaining about the amount of labour required. In 1774, he was complaining of insufficient salary, and the following year he requested extra labour. Eventually he, like Miller, was permitted to apply a little personal entrepreneurship to make ends meet and here his canny side shone through. First, he took issue with his accommodation, under threat, he reckoned, from the dangerous flues of the adjacent greenhouse. Could he have an extra £25 a year to supplement living elsewhere, and not over the shop, so to speak? At the same time, he pointed out that the salaries of himself and his labourers had remained static for several years, and inflation had eaten into the real value. Then, in the neighbouring Paradise Row, now Royal Hospital Row, he opened a shop to sell superfluous plants, raised in an agreed portion of the garden, all of which he achieved with the Society's blessing.

Forsyth never travelled, as far as anyone can ascertain, but his growing expertise and tenacity, which proved he could nurture plants from foreign lands, was the vital link in the survival and development of the plants he received. It was one thing for botanists to collect, press and bring home specimens, then study and place plants in the new Linnean system of classification which Forsyth adopted with alacrity. Quite another to take a scrap of root or seed of alien origin and coax a healthy plant to grow from it, which the public might then buy in substantial quantities. Endlessly strapped for cash, the Society wanted a gardener to garden, but they also wanted a man of high skill and intelligence to come up with the commercial goods, on a regular basis.

Forsyth had just the right combination of qualities. He described once finding that two pots of 'Barcelona' walnuts which he had sown some distance apart within the cold frame had been stolen by mice in one night, and hoarded by them for eating in the winter. By dint of searching round the pots kept in the frame he rediscovered the nuts and replanted them with the added security of placing slates on top. He went on to produce successful trees.

Forsyth became aware that because the fires had to be kept burning in the hothouses day and night over severe winter weather, leaves became scorched in the dry atmosphere, and fell off like deciduous trees in autumn, producing, as he put it, 'a very disagreeable appearance'. At eight in the morning, with the promise of a sunny day, his solution was to flood the tiled floor of the greenhouse with water, only opening the vents if the temperature reached 80°F. Around noon the water had evaporated and the floor was dry. By repeating this exercise two or three times a week in similar weather conditions, the plants produced leaves.

On 18 June 1784, he resigned from the Chelsea Physic Garden to become the gardener to King George III at Kensington. Unlike poor old Miller, he was still in the prime of his gardening life, and was in such glowing good odour with the Society according to *The Apothecaries' Garden* that he 'rec'd the thanks of the Committee for his great care of the Garden while in the Company's service'. Off he went not only to be the king's gardener, but to cash in on his last twenty years' experience. Not for him, though, a gardening manual suitable for the back gardens of artisan Britain. After producing *Observations on the Diseases, Defects and Injuries in all Kinds of Fruit and Forest Trees* in 1791, he moved to exploit further his reputation as being a thoroughly safe hand with fruit and vegetables with *A Treatise on the Culture and Management of Fruit-Trees* in 1802, which went into three editions. Alongside directives on the management of apples, raspberries, pears, cherries and nuts of many types, he also gave copious directions on cultivating vines, figs, peaches and nectarines. Glasshouses were confined to the very wealthy, but folk could always dream of fruits beyond their ken. Forsyth understood how to inject a little of the armchair horticultural adventure into his book.

Perhaps he gleaned a clue to commercial success from the initial literary flop of the man who had been his superior at the Society's garden in Chelsea, whose role bore the title Praefectus Horti. William Curtis produced a heavily illustrated document, *Flora Londinensis*, a detailed account of all plants growing within a ten-mile radius of the then centre of London, over which he had laboured for ten years. Few wanted to subscribe to an expensive selection of illustrations – 5s for coloured, and 2s 6d for black and white, especially as they could probably view many of these plants within a few minutes' walk. Appetites whetted by the arrival of plants from all over the world, the general public wanted to see and know about those which came from faraway places. Curtis went on to publish the *Botanical Magazine*, three plates of exotic plants for 1s, and his success lives on as the magazine continues to be published to this day.

Forsyth, in the meantime, continued to flourish as the head gardener to George III, and his expertise with fruit and vegetables took him to dizzying new heights of both fame and small fortune. Many of the king's fruit trees required severe pruning to remove the cankerous growths, and the raw wounds simply opened up the trees to disease. How he deduced his remedy is lost, but Forsyth's recipe remains. He mixed up cow dung, lime and wood ashes and bound the lot together with sand. This he plastered on the open wounds, where there must have been some good effect, even if it was impossible to detect if this was due to his mixture or natural resilience and regeneration. The pinnacle of his fame or infamy, depending on whom you asked, however, came a few years later, and was owed loosely to the success of Napoleon.

British ships were traditionally built of oak, and the requirement for more ships meant more oaks. The royal oak forests which supplied these trees were in a parlous state, the trees being old, fragile, and diseased. Mention was made of the 'plaister' of which Forsyth appeared to hold the secret to the quantities within the recipe. This might do the trick, cure the oaks and help to save plucky Great Britain against 'Boney' or the 'Bogey man' Napoleon. In true British style a committee was immediately formed, drawn from both Houses of Parliament, to consider the merits of Forsyth's 'plaister'. Approval followed, and he was granted £1,500, which was an enormous sum, for a mixture of relatively inexpensive ingredients, but desperate measures called for desperate remedies. The mixture proved at best controversial and at worst it was accused of being worse than useless. But the wily Forsyth rode the storm. He was the man whose hands had nurtured countless exotic plants; who had walked from Aberdeenshire through the gardens of Syon and the Chelsea Physic Garden; who had planted the seeds that produced flowers, fruit and vegetables for the king; and who was a talented and successful author.

Philip Miller is remembered only by one of the daisy-type family of flowers, *Milleria quinquefolia*. In contrast, William Forsyth has an unquenchable, long-lived place in gardens old and new, of the wealthy and poor, in the form of the reliable shrub *Forsythia* whose brilliant yellow flowers herald spring as surely as drifts of daffodils.

William Forsyth had come a long way from his humble origins as an under gardener in the rural heartland of Aberdeenshire, both in miles and achievement. He probably never returned to his roots, but he kept up the tradition of employing Scots, just like Miller.

Without Miller, Forsyth and Aiton (who was by the 1770s head gardener at Kew), whose hands nurtured the flood of plants arriving from abroad, and whose success quickened the desire for more and more 'new' garden exotics, there might never have been the vital link between Scottish gardeners and Scottish plant collectors. They were the directors, producers and stage hands who kept the plant show on the road, and their latest show was to burst open as surely as if a rainbow had exploded scattering flowers the colours of jewels in a landscape which was all too often muted.

Quiet, self-effacing, sandy-haired, modest Francis Masson was the man who made all this possible. In 1912, the local Aberdeen paper was famed for running the headline 'Aberdeen man lost at sea' above the story of the sinking of the *Titanic*. For Masson they might have led with, 'Aberdeen man returns with curious flower from African sojourn.'

CHAPTER 3

Francis Masson (1741–1805)

AN ABERDONIAN EYE FOR THE EXOTIC

Not only did Francis Masson's life span run parallel to William Forsyth's, it is likely that they were born no more than twenty miles apart. It is also very likely that they would have spoken the same dialect, the Doric, which although English- rather than Gaelic-based, would have been virtually unintelligible to those south of Edinburgh. Over two centuries later this still holds true. Perhaps it was this confusing way of speech that gave rise to the strange rumour that Masson was descended from Huguenot stock. Even fifty years after his death, and also that of Forsyth, an article in the *Cottage Gardener* of 1852 refers to his being of 'French extraction'. However, there is not one whit of evidence that Masson was anything other than an Aberdeen-born Scot. The surname Masson is one of the most common in the Aberdeen area. However, he was shy, diffident and a reluctant correspondent, and some may well have tossed in a whiff of French as an easy route to spicing up his biography. It may also have sat uneasily with his biographers after his death that, given a life of such contrasts, events and achievements, he came from a simple background, and remained diffident and modest to the end of his days.

In reality, Masson seems to have been a city boy from Aberdeen, in North-east Scotland, which in his youth was a settlement of only twenty small streets and 10,000 residents, but boasting two universities. Far from being a backwater, Aberdeen had long been a city of enterprise and culture. Today, the city still has two universities, and has grown twenty-fold. Masson would have received good primary education, and almost certainly served an apprenticeship as a gardener. In 1759, the year that he decided to seek his fortune elsewhere, there was an auction of books at Marischal College, the seat of one of Aberdeen's universities. Although possibly only around 20 per cent of Aberdonians were literate, the auction, as advertised in the *Aberdeen Journal,* was of 'Hebrew, Spanish and Arabic books, pamphlets and maps'.

Within the same issue there was a large advertisement aimed at recruiting soldiers for King George's 'third Regiment of foot guards, commanded by the right Hon John, Earl of Rothes, and in the Hon Captain Cosmo Gordon's company'. Masson apparently did not relish taking the king's shilling and joining the local regiment. With the same paper offering hints on avoiding starvation, it couldn't have been the most tempting of times to be an agricultural worker. All in all, Masson must have found a career in the area less than appealing. At around twenty years old, he set out on the 400-mile journey to London.

Whatever we don't know about Masson's life, we do know that royalty played a powerful off-stage role. The quiet Masson was not only a man of tough physical powers, outstanding as a naturalist and botanist, he was in the right royal place at the right royal time. The next we hear of him he was installed as one of the king's gardeners at Kew, and was just thirty when George III came to the throne in 1760 taking over Kew, both gardens and house. As the king was sizing up the gardens and gathering the reins of kingship, two of his most adventurous subjects were planning a round-the-world trip that would change Masson's life for ever.

Captain James Cook, the most famous adventurer of his day, was hoping to consolidate his role as circumnavigator par excellence and Sir Joseph Banks, already a seasoned traveller, had been aboard the

Endeavour with Captain Cook on a previous circumnavigation of the globe. Not only was Banks a giant in the field of exploration and scientific advancement of the age, he was also extremely wealthy, influential, confident, involved in international affairs and a man of letters. Ambitious and wily, he was well acquainted with the powerful in the land, and even with the king himself. Banks was aiming for the South African Cape, which had already yielded up desirable treasures for the great gardens of Europe. Since the early part of the eighteenth century, the Dutch settlers, and certainly their Governor Wilhelm van der Stel, had enchanted their homeland with *Pelargonium zonale*, with its exquisitely patterned leaf. The scarlet *Pelargonium inquinans* was added to this and a winning formula for summer bedding schemes was born. Banks himself had seen enough of South Africa's botanical temptations when he had stopped off there with Cook on his previous trip, and of what other delights were still there for the taking, he had more than an inkling. He was determined that Britain should not be usurped by a keen Dutch botanist or the Swede Carl Peter Thunberg, who was making a name for himself.

Originally, the wealthy Banks, whose three-year round-the-world trip with Captain Cook had produced vast quantities of dried plants from Australia and New Zealand, as well as a tantalising glimpse of the area round Cape Town, had been angling to accompany Cook on the second trip. Heavyweight, wealthy and capable of influencing the king, Banks forced his case, wheedled, bullied and demanded, but finally to no avail. The Admiralty was uncooperative. Among the many disagreements was Banks' wish to be accompanied by his pack of greyhounds and his personal orchestra on a Royal Navy ship. But he was not content to be denied.

In Banks' world, wealth counted; it opened doors, and he was used to getting his own way. Botany and natural history were his obsessions and he promptly turned to making sure he could send someone who could action at least part of his thwarted plan to scoop up the botanical treasures of South Africa. Francis

Masson appeared the ideal replacement: he was well trained in the cultivation of tender species, and was of sturdy physique.

Masson, the quiet Aberdonian, was therefore plucked from his secure job as under gardener at Kew and dispatched. Banks, as the man in overall charge of the king's gardens at Kew, pressed his head gardener, William Aiton, insisting that Masson was the ideal man for the job. He could be trusted to explore and safely bring back specimens which were known to be growing there. Once the expedition ship *Resolution* was off and away, Banks consoled himself with a trip to Iceland. Chartering a boat and crew, for which he himself paid, he sailed off for a slightly chillier type of botanical exploration.

As the *Resolution* would be away for many years, it was imperative that it was operationally first class. The Admiralty dithered about the cost of explorations. They certainly did not like spending cash on fitting out a mere exploration vessel, as it was not, in their eyes, as deserving of being in tiptop shape as the warships which were to stand between the shores of Britain and the perceived hooligans of the era, such as the Dutch or French who waited with bared teeth just across the Channel. They had a point: war with one or the other was to dominate the Royal Navy in years to come. Even if the exploration ships were doing useful tasks like claiming the odd highly strategic island for the king, or indeed a whole corner of a continent, this was of little use if the Navy were in desperate need of ships and some of them were at least a year's sailing time away and difficult to contact. It was anyone's guess if they would be still afloat.

On leaving Deptford on 9 April 1772, the *Resolution* meandered as far as Woolwich, only a few miles down the estuary, when she was hemmed in by easterly winds. The expedition then limped down the English Channel, stopping off to make some alterations to her superstructure, before finally setting off for the open seas accompanied by the *Adventure*. All went well from there, and the ships finally dropped anchor at Table Bay on Friday, 30 October.

By December, Masson had searched out two companions and

was ready to set off. He referred to one of these men as a Dutchman, but he was actually a Swede, Franz Pehr Oldenburg. Oldenburg had been in the Dutch East India Company as well as being a private soldier or more probably a mercenary. It was possibly the reference to the Dutch East India Company that confused Masson. Also accompanying him was a man whom he described as a 'Hottentot', the accepted descriptive term at the time for a native black South African, who drove the wagon, which was pulled by eight oxen.

On the evening of 10 December they left Cape Town, travelling to the north-east. Masson was released into a land about to burst into flower. However, he soon found that his challenges were not to be confined to searching for plants: his fellow travellers, and other plant searchers, were to be his main worry. This strange little trio, equipped with a wagon in its last stages of deterioration, crossed the Great Berg River and entered the Drakenstein Valley. Masson was most impressed by the vineyards and orchards of European fruits which had been transported 'by the Dutch; viz. apricots, peaches, plums, apples, pears, figs, mulberries, almonds, chestnuts and walnuts, but no Indian fruits, except the guava and jambo, neither of which ripen well'. But they also came across French refugees, Huguenots, who eked out a living on poor soil, but still managed to produce 'some wine and summer fruits'.

As a highly inquisitive gardener, Masson had a well-honed eye, curious to know what grew. He also had a sharp eye for which plants might flourish in Britain, even under glass. Masson was impressed by the irrigation system. He described plantations at the foot of the mountains which were watered by swift, small streams which cascaded down to the gardens and vineyards. In the next village they found a dozen houses, with pretty gardens and vineyards that produced 'excellent wine'. The next day he took to hunting the local antelope called 'Ree buck' (roebuck), but with no success. He had already seen huge numbers of zebras, which he noted were called wild horses by the local Dutch inhabitants.

Like many explorers before and after him, he was to note that the everyday life around him was just as fascinating as curious new plants. As they lumbered along on their wagon, Masson described in chatty style the local goings-on.

> We reached Stellenbosch, a small village about 30 miles north-east from Cape Town, consisting of about 30 houses, forming one regular street, with rows of large oak trees on each side along the front of the houses, which render it very pleasant in the hot season. These oaks, of the same sort as ours in England, were brought out of Europe by Adrian Vanderstell, former Governor of the Cape, who built the village and gave it his name. The country around it is populous, and contains many rich farms, which produce plenty of corn and wine. The farmers we found busy in treading out their corn; which is performed by horses in the following manner. They make a circular hut of about 30, 40, or 50 feet in diameter, with the composition of clay and cow dung, which combines very hard; round it they erect the mud wall, about breast high; this floor they cover with sheaves, beginning in the middle, and laying them in concentric circles till they reached the outside. They then turn in about 20 or 30 horses, which a Hottentot, furnished with a long whip, drives round and round till the corn be trodden out, and the straw becomes as fine as chaff; which they afterwards clean, and carry into their granaries. This method they can practise with great security, as it seldom rains from the middle of October to the middle of March.

They proceeded on to the mountains, passing the eastern part of False Bay, and finally, taking 'an extremely rugged steep path, and after much labour and fatigue, gained the summit'. So strange did the mountains appear that he described the rocks looking like the ruins of church steeples, but at last he was able to report that there, in the area where the soil about them was black mixed with pure white sand, was a region which he thought might be 'the richest mountains in Africa for a botanist'. After the excitement of this day, now quite confident that a rich harvest might be there for the picking, they lodged at a 'mean cottage', where he was shocked enough to note that the Dutchmen, who were the farmers, slept at one end, and their servants, the Hottentots, lived almost 'promiscuously together, their beds consisting only of sheepskins'. They then proceeded in a north-easterly direction to the Fransch Hoek Valley.

The journey wasn't without incident. Trembling and dazed by an encounter with wolves, most probably only jackals, Masson was still able to write:

> We afterwards climbed over many dreadful precipices till we arrived at the woods; which are dark and gloomy, interspersed with climbing shrubs of various kinds. The trees are very high; some from 80 to 100 feet; often growing out of perpendicular rocks where no earth can be seen. The water sometimes falls in cascades over rocks 200 feet perpendicular, with awful noise. I endured this day much fatigue in these sequestered and unfrequented woods, with a mixture of horror and admiration. The greatest part of the trees that grow here are unknown to a botanist. Some I found in flower; others which were not so, I'm obliged to leave[.] [T]he researches of those who wait may come after me[,] in a more fortunate season.

He was worried already that he might be missing out, visiting in the wrong season. Within a few days they thankfully arrived at a hot bath or spa situated on the south-east side of the Zwartberg mountains, later known as the Caledon Baths. After calling at the Dutch East India Company depot where they rested for five days, they then returned more or less by the same route. Masson realised that what he had seen was only a tantalising glimpse of the many treasures which lay around.

He had also found a new associate, Carl Peter Thunberg, who was formerly a pupil of Linnaeus in Sweden. They seemed unlikely travelling companions. Thunberg was cocky and full of his own importance – as a qualified medical doctor and a swashbuckling adventurer who relished danger, he certainly considered himself more than a cut above Masson. He also considered himself a natural-born leader. In reality, he was a scientist, Masson a gardener, and together, even if Thunberg was loath to admit it, they had an ideal partnership. Thunberg would certainly have thought his knowledge greatly superior to Masson's, a mere hands-on gardener. However, it was Masson's observant eye that would pick out species even when not in flower. Thunberg would then provide the more scientific and botanical analysis as to their precise species. The practical Masson and the more esoteric Thunberg were a complementary and equal partnership.

Thunberg was quick to note that his fellow traveller was 'an English gardener, who had been sent hither by the King of England to collect all sorts of African plants. He was well equipped with a large and strong wagon tilted [furnished] with sailcloth, driven by European servants upon whom he could depend.' Together they travelled west, and although each contributed differently to the plant-finding partnership, their skills at survival and adventuring were more or less on par. Thunberg, of course, considered himself the leader of the expedition, but in fact, it was perhaps only through luck that the pair survived at all.

It was the rainy season and the expedition had spent the night at a farmhouse. In the morning, the farmer's wife had sent one of her servants to point out where they were safest to ford the river, now a raging torrent in place of the normal trickle. Language difficulties intervened and, as they made ready to enter the river at the most dangerous point instead, Thunberg waxed eloquent:

> I who was the most courageous of any of the company, and in the course of the journey, was constantly obliged to go on before and head them, now also, without a moment's consideration, rode plump [sic] into the river, till, in a moment, I sunk with my horse into a large and deep sea horse [hippo] hole, up to my ears. This would undoubtedly have proved my grave, if my horse had not been able to swim; and I, who have always had the fortune to possess myself in the greatest dangers, had not, with the greatest calmness and composure, guided the animal (which foundered about violently in the water) and kept myself fast in the saddle, although constantly lifted up by the stream. Holes of this type which the hippopotamus treads out for its resting place, occur in a great many rivers, though the animal itself is no longer to be found there as it has either been shot, or made to fly to some more secure

retreat. All this time my fellow travellers stood frightened on the opposite bank and astonished, without daring to trust themselves to an element that appeared to them so full of danger. However, as soon as I had got off my horse and let the water drain off me a little, I ordered my servant to drive across the river, according to a better direction that I gave them, after which the others followed.

Clearly, they already knew quite a lot about hippos by this stage, although they had observed almost none, as the Boer farmers had slaughtered so many for food. The quality and quantity of meat obtainable from a single animal was very attractive in a harsh life. Masson, on the other hand, gave a pretty terse account of what he concluded was a foolhardy dash into an obviously dangerous situation:

> The doctor imprudently took the ford without the least inquiry; when on a sudden, he and his horse plunged over head and ears into a pit, that had been made by the Hippopotamus amphibius which formerly inhabited these rivers. The pit was very deep, and steep on all sides, which made my companion's fate uncertain for a few minutes: but after several strong exertions, the horse gained the opposite side with his rider.

Soon they came to an area which Masson called Carro, probably part of Little Karoo, and here he was astonished to see 'within an area of a reddish colour intermixed with rotten rock . . . amazing evergreen shrubs'. From here he gained *Cotyledon*, *Euphorbia* (spurge), *Portulaca* (sun plant), and *Mesembryanthemum crassula*, which later produced the popular bedding plants, Livingstone daisies. Here too Masson saw his first lion, which fortunately kept a distance. He described the local buffalo as being so much larger and fiercer than our British oxen. But he cautiously observed that 'In the daytime they retired to the shelter of the woods, which renders it very dangerous to botanize there.'

Despite these dangers, treasures awaited around the corner. Crossing the Van Staadens River, they trekked on ground carpeted with the flowers of which they had dreamt. *Protea*, which became the flower emblem of South Africa, were gathered, along with *Ixia viridiflora, I. cinnamomea, Erythrina corallodendron* and *Ornithogalum thyrsoides,* a white bluebell-like flower prettily named chincherinchee. They returned, half starved but triumphant, to Cape Town. They may have felt as though all they desired was to stay put and take time to box up their botanical finds and dispatch them straight to Europe, but they had scant chance of that. Almost immediately they departed again, exploring less arduous tracks with an adventurous, talented, knowledgeable and indomitable army wife, Lady Anne Monson, a great-granddaughter of Charles II. Although already sixty, she was an intrepid botanist and was at ease wandering around the surrounding hills with two men thirty years her junior. She was also a scholar of such repute that Linnaeus named the genus *Monsonia* after her. She took both Masson and Thunberg firmly under her wing. Thunberg wrote slightly breathlessly about her that not only had 'this learned lady, during the times she stayed here, made several very fine collections, and particularly in the animal Kingdom . . . I frequently had the pleasure, together with Mr Masson, of accompanying her to the adjacent farms, and contributing greatly to the enlargement of her collections, and she had the goodness, before her departure, to make me a present of a valuable ring, in remembrance of her.' Lady Anne, a memsahib in the making, proceeded on to India. She herself left no record of her encounter with the intrepid pair.

Despite having had more than a few tense moments together in their first foray, Masson and Thunberg then undertook a second expedition in tandem. For Masson, this was to be his third major adventure, and he was worrying about costs. His salary may have been paid by the king, but it was exceedingly modest. He had also set up a garden in Cape Town and was paying somebody to look after it so that the seeds and plants that he found were well established in cultivation before they were sent home. Ever the practical man, he wanted to observe their growing cycles, which from experience of gardening at Kew he knew would be critical to both their survival and his good name.

He was also reflecting on the way of life in this unusual country. His concern and reaction to various human rights violations were typical of the many plant collectors who followed his example in other parts of the globe. Neither Masson, nor any of his companions, were military men or colonisers. They were merely there to gather seeds and plants, and when they moved on they left little trace of their picking and digging. They were totally dependent on the goodwill of the local people. They had no wish to exploit anybody, nor indulge in wholesale plundering. The very nature of gardening imbued them with patience and powers of sensitive observation in huge quantities. There was little point in raging at a plant that refused to flourish. Self-control and acceptance of the vagaries of plant life produced a respect for all nature. Masson was a gentle soul, and the slaughter of wildlife distressed him. By the time he arrived in the country, the Boers had almost denuded the area of hippos, which they regarded as an imperfect but acceptable substitute for pork.

Perhaps his botanical background had also given him more than an inkling of the food chain, and of how fragile botanical and animal life might be in the face of man's harsh impact. His knowledge of plants would have made him all too aware of just how vital insect and bird life was to successful pollination. Perhaps he would have been more than able to deduce that such wholesale annihilation of the local animal life was not a good thing. However, much worse was his horror at the way many of the local, indigenous people were treated.

Masson was in this, as in many other ways, a man before his time. But while his natural reticence, lack of self-confidence due to a restricted education, and modest background restrained him from voicing such thoughts aloud, his conscience ensured that he committed them to his diary. He was aghast at the poverty of the local native peoples, and noticed that their lands and way of life had been decimated by the Europeans. 'A few Hottentots still remain here, who live in their ancient manner; but are miserable wretches, having hardly any stock of cattle,' he observed unhappily. The term

Hottentots was used for indigenous African peoples who had been pastoralists, as opposed to the hunter-gatherers. The South Africa in which Masson arrived in the 1770s contained fewer than 14,000 'burghers', or middle-class émigré Europeans who sought a better way of life, consisting of Huguenots who had fled from France to the Netherlands in 1685 after the French government decreed Protestants unwelcome. Huguenots fled not only to South Africa, with its tolerant Protestant Dutch government, but also to England and North America. But there was a great contrast between the high calibre of the large numbers of Europeans settling in America at the same time, who were skilled workmen, professionals and minor aristocracy, and the majority of whom showed astonishing determination and resilience, and those arriving in South Africa. The latter generally came from amongst the less well educated, often from insecure financial backgrounds, and more often than not sought simply a soft option. In addition the Dutch East India Company had traditionally paid poorly and rarely attracted employees of high calibre; only the desperate joined as the dangers in the countries in which the Company operated were well known.

Into this hotchpotch were added slaves, often seized from the Indian subcontinent. Their numbers just matched the Europeans. Masson was appalled to come across a party of Dutchmen out hunting slaves with every intention of killing them, as they held them responsible for stealing their cattle. The execution of slaves was often preceded by barbaric torture and dismembering. As if to drive home the distinction between the native peoples, coloured slaves and European settlers, two sets of gallows were situated outside of the main town: one for Europeans and the other for slaves and native peoples.

Masson himself had already come within a whisker of a collection of Hottentot convicts who had escaped into the mountains. Only by sheltering, hardly daring to breathe, for one terrifying night in a hut he stumbled across, did he manage to evade them. However, despite his dangerous experiences, and in line with his enlightened views, on the third of his treks into the interior, Masson set off in the sole company of two 'Hottentots'. Thunberg joined him a week later.

For this journey Thunberg had upgraded his transport and replaced the old wreck they had had the year before. He was now the proud possessor of a brand-new cart, covered with sailcloth, of which he was quite proud. Thunberg had a clear sense of drama, and always wrote with an eye towards the history books, in which he felt certain to be included in large measure. He described the expedition thus: 'We had each of us a saddle horse, and for our wagon we had several pair of oxen. Thus we formed a strange Society, consisting of three Europeans and four Hottentots, and for the space of several months were to penetrate into the country together, put up with whatever we should find, whether good or bad, and frequently secluded ourselves from almost all the rest of the world, and of the human race.'

They progressed north, passing close to present day Malmsbury, over to Saldanha Bay, where Masson found a great variety of curious plants, one of which was a species of showy amaryllis, which, he observed, 'the local Hottentots use to poison their arrows for the shooting of small game'. Later the Europeans used it as an antiseptic, using the scales dipped in oil and water and placing these coverings on wounds.

The expedition travelled north as far as Olifants River Valley, passing by some hot springs where he observed orange trees growing happily out of a seam in a rock, with their roots almost in boiling water. Olifants River had been named by an adventuring seaman, Jan Danskaert, who in 1660 left the Cape Town area to explore. The story goes that he named the area after the thousands of elephants he saw in the area. Elephants became Olifants, and although Masson never mentioned seeing elephants, he did note the area for its rich variety of vegetation, and potentially excellent growing conditions. He was correct and the region was to become a major wine-producing area within a few years of his visit. Its fame spread at least 2,000 miles, to St Helena, where the exiled Napoleon, no doubt with time on his hands, somehow got wind of the fact that the dessert wine was of a high quality, and commanded that some should be brought to his remote island. No doubt rather pleased to have secured a captive market, so to speak, for their wine, the vineyard owners were only too pleased to heed the command.

This part of the expedition presented new obstacles. As it was high summer, in November, the sandy country towards the mouth of the Olifants River was not only extremely hot, but the ground itself was unstable owing to the huge numbers of dune mole rats, or sand moles. As they tunnelled, they created a fragile top layer of ground. The horses stumbled constantly as their weight crumbled the ground – their hooves easily shattered the top crust of the earth and they frequently sank up to their shoulders in sand. Masson observed the mole rats feeding off the very bulbs that he was collecting, such as *Ixia* (African corn lily) and *Gladioli* which, though beloved by many, were sadly sneered at by gardening arbiters of good taste by the middle of the twentieth century, as their spiky, unfolding flowers appealed to suburban gardeners fond of rigidly straight lines.

Masson and Thunberg proceeded north as far as present-day Nieuwoudtville before turning and wandering down through the Roggeveld mountains, coming south as far as present-day Michell's Pass near Ceres, descending to Swellendam, before moving along to Mossel Bay. They now progressed as far as Algoa Bay along a coastline that is today acclaimed as the Coastal Garden Route and organised for tourism. However, the Coastal Garden Route that Masson and Thunberg rode through was in its original, virgin state. Local farmers would have been intent on growing crops for survival, rather than eyeing up the carpets of flowers as a cash product. Masson and Thunberg were lucky to have any bed under cover apart from their wagons. Occasionally, they struck lucky and spent the night with a farmer.

Hendrik Swellengrebel, the son of a former governor of the colony, left behind a description of the living conditions of the farmers, written the year after Masson's departure, in 1776–77.

In the area in which Masson and Thunberg had travelled, there were some

> quite respectable houses with a large room partitioned into two or three areas, and with good doors and windows, though mostly without ceilings, open to rush roofs. For the rest, however, and especially those at a greater distance in more remote areas, they are only tumbledown barns, 40 feet by 14 or 15 feet, with clay walls four feet high, and a thatched roof. They are mostly undivided; the doors are reed mats; a square hole serves as a window. The fireplace is a hole in the floor, which is usually made of clay and cow dung. There is no chimney; merely a hole in the roof to let smoke out. The beds are separated by a Hottentot reed mat. The furniture is in keeping.
>
> I have found up to three households including children living together in such a dwelling. The majority, by far, of the farmers from beyond the mountains come to Cape Town only once a year, because of the great distance – I have discovered that some of them are reckoned to live 40 days' journey away – and because of the difficulty of getting through the passes between the mountains. To cross them they need at least 24 oxen, two teams of 10 to be changed at every halt and at least four spares to replace animals that are crippled or fall prey to Lions. Two Hottentots are necessary as well as the farmer himself. The load usually consists of two barrels of butter weighing 1000 pounds in all, and 400 to 500 pounds of soap.

Masson took every opportunity to learn, botanical or not, during his three treks into the interior, and his contact with the farmers of the Coastal Garden Route was no exception. Stopping for a rest for several days with an earthy Dutch farmer, he revised his low opinions of the Dutch 'peasant' farmers he had encountered. Too many times he had had cause to be appalled at their treatment of their newly acquired land. But this encounter was very different indeed. Masson was greatly heartened when the man revealed to him 'that in winter the hills were painted with all kinds of colours', and said it grieved him often that 'no person of knowledge in botany had ever had an opportunity of seeing his country in the flowery season'. Furthermore, when Masson expressed astonishment at the quantities of sheep, he found out that they never ate grass, but only succulent plants and shrubs, 'many of which are aromatic and gave their flesh an excellent flavour'. Within a day he watched this in action, as the sheep munched their way through even the seedpods of *Euphorbias*. By now Masson and Thunberg could only concentrate on collecting within a few metres of the roadside, as their animals were desperately thin and starving, so it was hardly surprising that this encounter with a man who appeared to have sympathy and appreciation, as well as having adapted his way of life to the area, made such an impression. At this difficult stage in the expedition, Masson's growing respect for the native Africans' knowledge and exploitation of the natural world emerged constantly in his diary. To this he added the Dutch farmer.

A new species of aloe, he observed, *Aloe dichotoma*, was used for making quivers for their arrows. The arrows had tips poisoned with 'the venom of serpents mix[ed] with the juice of a species of *euphorbia*'. They had already had close encounters with snakes. As they trod clumsily over the rocky, stony ground, snakes would be startled and rush off, even crawling between the horses' legs. When Thunberg, Masson and their servants rested, on at least one occasion snakes slithered over their bodies, and one actually entwined itself round Masson's leg. But the reptiles appeared to be placid and, as neither man mentions it in his accounts, it appears that no one was bitten.

After a loop inland to Little Karoo, Masson and Thunberg arrived safely back in Cape Town on 29 December 1775, and Masson sailed for England in the late spring of the following year. Years later, William Forsyth, during his time as the head gardener at the Chelsea Physic Garden, rewrote Masson's adventures, embellishing many of his adventures to new dramatic heights, and his second-hand account of his associate and friend lost nothing in the telling. He stated, for example, that 'Their "African servants" had refused to allow them to advance into land occupied by the Kaffirs [a term used at the time for some native South Africans] as they would have been murdered for sure, if only for the iron on their wagons.' These are descriptive and imaginative embellishments from a suburban gardener who, for most of his adult life, had strayed no more than a couple of miles from London.

Back at Kew, Masson was greeted by Sir Joseph Banks with relief that he had returned alive, and pleasure that he had brought back a treasure trove of plants. Banks' decision to persuade the king to fund the trip had paid off handsomely, and Banks basked in reflected glory.

Much had changed on the world stage in his three-year absence. The colonies of America had successfully broken away from the jurisdiction of Britain. George III had lost much, but perhaps he could derive a smidgeon of comfort from the exciting new flowers for his garden and glasshouses. The king was delighted, and Banks glowed, confirmed in his status as a man with great foresight, who would organise scientific collection from 'new' continents with success, heading off the efforts of rival countries. Thunberg certainly returned to Sweden with bountiful collections of seeds, similar of course to those gathered by Masson. But whereas Thunberg was more of a collector, and observer, the ever practical gardener Masson had had the advantage of being able to grow his plants in his garden in Cape Town, before shipping them home. There, they had the double advantage of being tenderly and knowledgeably cared for by Forsyth among others, whose skills in caring for tropical plants had been honed in the greenhouses of the Chelsea Physic Garden. The king was pleased with Masson and Banks, and gave Banks a generous reward.

William Aiton, the canny Lanarkshire gardener in charge at Kew, on whose recommendation Masson had gone in the first place, greeted him with understated professional acclaim and took great advantage of the arrival of his collections. Basking also in the reflected glory of personally recommending Masson for the trip to South Africa, Aiton not only applied his skills to nurturing the tender plants sent back by Masson, but commenced on his great volumes of *Hortus Kewensis or, a catalogue of the plants cultivated in the Royal Botanic Garden at Kew*, which detailed every plant received there in his tenure. And although many of the plants were already known to exist, James Edward Smith, founder of the Linnean Society, pointed out that 'but comparatively few of them had been procured in a living state, or cultivated with success, even by the Dutch themselves, and of those but a very small portion had, from the time of the First Earl of Portland come into general cultivation'.

Years later, Smith was to deduce that the 'establishment of a travelling botanist in the King's service, if not suggested by the first mentioned [Aiton], of these eminent men [William Aiton, Sir Joseph Banks and Daniel Solander] was planned entirely under his advice and direction'.

For Masson himself, South Africa had been providential. His finds were to stun the horticulturists back in Britain. If he had sent back birds such as toucans and parrots of startling shapes and hues, which then set up home amid the British trees that normally played host to conservative sparrows, wrens, blackbirds and pigeons, the effect could not have been greater.

Contract agreements were honoured all round. As arranged, Masson received his £100 per year for each year he had been absent. His expenses totalled £583 8s 6d, a sum happily just under the £600 living expenses which he had been allowed. Back home in Aberdeen, his gardener's wage would perhaps have just topped twenty pounds a year, with the free house, meals, grass for grazing, and a token amount of produce from the laird's garden surplus. All in all, as a first expedition, it had been an astonishing success.

Some idea of the transformative impact that Masson's expeditions had upon the British garden can be gained from Smith's accolade to his achievements. Writing in 1819, he recalls the pleasure he derived in 1779, just four years after Masson arrived home, from 'the novel sight of an African Geranium [*Pelargonium*] in Yorkshire and Norfolk . . . Now every garret and cottage window is filled with numerous species of that beautiful tribe, and every greenhouse glows with innumerable bulbous plants and special heaths of the Cape'. Indeed, by 1864, the *Journal of the Cottage Gardener* was giving effusive hints concerning the propagation of *Pelargonium*, which by then had acquired names such as Cloth of Silver and Celestial, or been named after ladies such as Madame Sainton Dolby and Arabella Goddard – varieties now long forgotten. Commonly known as geraniums, these plants have evolved into scarlet, pink, white, multicoloured, frilly, double-flowered, single-flowered millions. They feature as potting plants in the window boxes of Perth in Scotland to Provence. They are synonymous with whitewashed Andalusian towns, and symbolise, mistakenly, as it turns out, the attractions of the sunny Mediterranean countries.

Masson, though, wanted more. He was disinterested in fame, and, luckily for him, content with modest remuneration. Masson still thought of himself as a simple fellow with a gardening background, and had no illusions about how the great and good regarded him. He did, however, harbour desires to be honoured by the person he saw as the most eminent botanist of his era. But for Masson to approach Linnaeus to request a species to be named after him was like approaching God.

In the meantime, the strutting Thunberg had raised his opinion about Masson. His respect was to aid Masson in his dream and it was Thunberg who apparently suggested to Linnaeus that a genus should be named after him. Masson maintained a curious interest in strange, unusual plants, few of which achieved the blanket fame of *Pelargonium*. A group of small low-growing plants were indeed given his name, *Massonia*. Plants of this group have two large leaves with a fragrant flower, white or pale mauve, consisting of dozens of upright stamens, rather like an upright washing-up brush. They do not easily flower in northern countries.

However, for a man who had returned with plants that flowered with zinging colours, London appeared a tad grey. He had spent three years under brilliant skies and hot sun, introduced such exotics as *Strelitzia reginae*, the bird of paradise flower, later named after Queen Charlotte, wife of George III. The flower looks just like a beaked bird with a strange blue- and orange-tinted hairdo, a quieter version of the mohican hairstyle worn by twentieth-century punks who sported vertical combs of hair, lacquered to defy gravity. This naming was perhaps unwittingly appropriate. Queen Charlotte was reputed at the time to have 'negroid' features, perhaps most ably reproduced by Ramsay, the Edinburgh portrait painter, for whom she sat several times. Ramsay and his father were strenuously anti-slavery, and it would have been quite in character for them to portray her accurately, whereas other portraitists would have thought better than to depict their queen's features so truthfully. Modern researchers have indeed uncovered her probable African genetic links, but here was a flamboyant flower from the heart of of South Africa's hinterland being named in her honour many years ago. That said, the naming of the flower probably had as much to do with ensuring that future trips would be funded as anything else.

After the brilliance and variety of Africa, of which the bird of paradise flower is but one example, Masson found it difficult to settle down again to gardening at Kew, albeit as one of the king's gardeners in the king's private garden. Masson felt stifled by the dull, strictly hierarchical routine at Kew and begged for another expedition. Banks again came to the rescue and within a few months Masson set off for the Canaries from whence he sent back to a grateful Banks in 1777 the *Senecio cruenta*, from which all the vast array of today's cinerarias are derived. Then he went off to Madeira, the Azores and the West Indies.

Still paid at the rate of £100 per year, his bills of passage, which itemised his berth, food and incidentals for travel, make fascinating reading when compared with travel costs today.

England to Madeira	£ 15 15s
Madeira to Azores	£ 10
Within the Azores	£ 30
Madeira to Tenerife	£ 4

Tenerife back to Madeira	£ 30	
Madeira to Barbados	£ 31	10s
Barbados to Granada	£ 3	
Granada to St Eustachia	£ 4	
St Eustachia to Antigua	£ 3	
St Eustachia to St Christopher	£ 3	
St Christopher to St Lucia	£ 6	
St Lucia to Nevis	£ 3	
Nevis to Jamaica	£ 10	
Jamaica to England	£ 40	
Totalling	£ 193	5s

Masson was being paid £100 per year, deemed a good wage for a government clerk in that era. A clergyman would have been paid around £182, and a barrister £242, whereas a lowly agricultural worker around £37, although he would also have been given a house and various perks. Putting this into context, the cost of his travel was colossal by our reckonings today. Even by using one of the comparable inflation tables, Masson's travel expenses at £193 were the equivalent of £14,500 in 2002. His trip from Jamaica to England at £40 equates to £3,250 in 2002, maybe acceptable for a luxurious cruise ship today, but for Masson's rudimentary accommodation, a hefty amount. No wonder that his employers were paying the minimum wage possible – their expenses were substantial.

But the journey was to cost Masson dear. Far from spending his time in the relative comfort of his own cabin on a ship, or at least in his very own caravan wagon in South Africa, this time he had been seized and flung into jail, having been captured by the French at Granada in 1779. Worse, Masson, the man who seemed unlikely to raise a gun except to hunt food, was forced for a brief few weeks to fight for his captors. He leaves virtually no accounts of this adventure, but as he appears rarely to have fired a gun, and noted many a time how he failed to hit a deer for dinner in Africa, he must have made a disappointing addition to the French forces. Somehow, he managed to wiggle out of the situation, and collect himself and his plants together again. But worse was to come. The war had disrupted shipping, and he could only sit and wait for a passage home while his seeds and plants rotted. Finally, when he must have thought that the situation could get no worse, it deteriorated dramatically. A hurricane hit St Lucia, the island on which he was waiting, and he lost virtually everything. The miracle is that he did manage to return with some tropical plants. His success, even in the wake of imprisonment, delay, hurricanes and loss, was considerable. Banks was satisfied, but, incredibly, Masson could think of nothing else but setting off again.

Banks lobbied the king mercilessly. Kew, he mentioned, was now vastly superior to every other similar establishment in Europe. He drew attention to a famous journey to the Levant by M. Tournefort, financed by the French king Louis XIV which had produced nothing like the treasures which Masson had brought back from the Cape. And furthermore, he added for good measure, pushing the financial angle and twisting the knife of the king's past mistakes as far in as he dared, 'As far as I am able to judge, His Majesty's choice of Mr Masson is to be accounted among the few Royal bounties which have not been in any degree misapplied.' Meanwhile Masson, while in London, kept up with many Scottish gardening associates. At the Vineyard nurseries in Hammersmith, he called on the firm of Lee and Kennedy. James Lee was by now an elderly man, and about to hand over to his son, also named James. This James had a partner Lewis Kennedy, who was to design the grandiose formal garden at Drummond Castle in Perthshire. Drummond Castle garden is one of the few ambitious formal gardens to survive complete until the present day; it was even enhanced in the early twentieth century. However, in the late eighteenth century, such formality was about to be swept away by the new relaxed style of English garden designers such as Capability Brown.

Masson, well acquainted with many types of garden design, found himself amid the formality of European gardening when he reached his next destination, Portugal. He laid out a garden in Lisbon, adding garden design to his skills. He sent back various species, and in a rare moment of criticism, reported that he had little time for Portuguese plantsmen. He moved on to Spain, visiting Cadiz, San Roque, Algeceras and Gibraltar before crossing over to Morocco. Oblivious to the well-founded rumours that Sale in Morocco was a pirate port, he sent plants gathered from close by there to Banks.

By the end of 1785, Masson was back at Kew, but hankering to return to South Africa. This was problematical as Britain was at war with Holland and her colony had imposed stringent security measures on most visiting foreigners, especially the British. As ever, Banks smoothed the way. Tucked away in his travelling trunk were credentials from the Dutch ambassador in London, Van Lijnden, and accreditation papers to the governor of the Cape Province from no less than the Marquis of Carmarthen, the British Minister for Foreign Affairs. Carmarthen requested that Masson be given permission to travel out from Cape Town to collect plants.

It was all to limited avail, as the governor, after weeks of deliberation, permitted him only to travel within agreed limits, which were roughly no more than three hours' journey from the coast inland. There were strong warnings from the Dutch that any foreigner breaking this embargo would be hauled in front of the governor for punishment.

To begin with, Masson behaved admirably and in his diary entries seemed fairly content: 'Since I last wrote I have had permission to visit the Hottentot Holland mountains for only five days and was so fortunate as to find some of the rarest *Erica* [heathers] and *Protea* [South African national flower emblem] in seed. I also found some new *Protea* which has not yet been fully described elsewhere.' He went on to send home 117 species, naming one species of *Erica banksia* after Banks. But he was frustrated and wrote home requesting a posting to India. When this was refused he decided to stretch his boundaries and if necessary suffer the consequences. He almost never even hinted at this in his letters, but while gathering together all the implements and personnel needed, he probably gave away to his masters back home that he was completely disregarding the orders. In addition to noting down bills for garden pots and boxes, stationery, baskets, powder and shot, he also put on his shopping list a large carriage and ten oxen plus a Dutch wagoner, cooking utensils and his various necessities for a journey. Then he billed for expenses, letting slip that this included a trip to the 'Elephant River' which was the Olifants area he had visited before and lay a good two hundred miles from Cape Town. In fact his expenses revealed his travels as clearly as though they had been drawn on a map.

Banks ticked him off in a letter which started off amicably enough, heaping praise on the quality of the plants he had sent home. Banks even went so far as to say that they were very much better than when he had been there on his first expedition. But why wasn't he searching out new species within the permitted areas of False Bay? Why didn't he just concentrate on this area until it was exhausted of finds? There was even a hint that he should 'remain quiet. Afterwards you may explore.' There was mention of sending him to Botany Bay in Australia, for plant collecting rather than as a convict, but Aiton managed to derail this suggestion by informing Banks, correctly, that Masson had never wished to be sent to Australia, and even had an objection to going.

Masson ignored it all and pressed on. His journeys of 400 miles up the north-west coast to Kamiesberg in 1789 were followed the next year by an inland foray to the Roggeveld mountain country to altitudes of 4,000 feet. This no doubt contributed to the pressing need for the purchase of a new wagon: this type of trekking, as the Dutch farmers had demonstrated ten years before, literally shook wagons to pieces. Despite the risks, Masson managed to evade the Dutch restrictions and returned with a multitude of plants in 1795. Banks, born in 1743, and Masson were now in their fifties, elderly by the standards of the age, and had known one another for twenty-three years. Banks was now one of the most influential and wealthy men of his day. He had also been elevated by a knighthood and as 'Sir Joseph' he was busy transforming Kew from just a pleasure ground into a centre for botanical reference.

Masson spent some time cataloguing and writing up his book about the curious *Stapelia novae* or star flower – a curious cactus-like plant with a star-like central flower – and loitering and languishing amid the greenhouses at Kew which stored his collections from South Africa. Within them was Masson's cycad, the *Encephalartos altensteinii*, which was destined to become the oldest known pot plant in the world. Brought back in 1775, it still is alive today, 230 years on. Did Masson look at this plant with its elephant-like stalk and think back to the use to which the Africans put the plant? The pith was scraped off the stalk, buried for a couple of months to rid it of its poison, and, when this time had safely passed, dug up, kneaded into a loaf, and placed amid hot embers to bake. Masson's other plants were blooming too. Meanwhile, William Aiton, who had rushed to catalogue the plants of the most comprehensive and wide-ranging botanical garden known, had died two years before. Aiton's son, William Townsend Aiton, was now in charge at Kew and spread his expertise over the royal gardens at Kensington, Buckingham Palace and – eventually – the Royal Pavilion at Brighton. But though the gardens at Kew were expanding, Masson felt confined.

Much of Masson's life had been spent free and unfettered in corners of the world that others could only imagine. Although now fifty-seven years old, he pleaded with Sir Joseph to intercede on his behalf with the king. He reiterated that he was in 'a reasonable share of health and vigour', although admitting he might be 'in the afternoon of his life'. He pointed out that he was more than able to continue exploring, and emphasised just how many areas of the globe were as yet untouched by plant collectors. If anyone could understand how he felt, Sir Joseph Banks could as his early life had been spent on one of the most extraordinary adventures in history, circumnavigating the earth with Captain Cook. Like Masson, Banks had survived many of those early explorers and he was certainly sympathetic and eager to help. It was sheer bad luck that His Majesty's Navy were preoccupied in various parts of the world. However, there was one country which was now, if not an ally, then at least in a state of mutual agreement. The American War of Independence was largely over and William Townsend Aiton and Sir Joseph Banks hatched a plan to send Masson to North America.

Banks penned a lengthy and effusive letter to the king; although he would have been well able to fund Masson himself, Banks would have felt it reflected greater glory for the king himself to fund the voyage, and Banks indeed wished to have the ear of the king. The king might have lost his colonies across the Atlantic, but perhaps it would be some recompense to bring back plants, if not from the embryonic United States of America, then from Canada which was at least on the same continent. Banks laid on his recommendation every persuasive word he could muster, ending his appeal to the king by bringing God to his aid. He corralled all his arguments, bringing to the king's notice that a fresh bag of plants would be a royal bounty 'which would conciliate the gratitude of all who make the science of nature their study throughout Europe, and more especially, those in this Kingdom, I may say, under his Majesty's particular auspices and protection, who follow that most engaging occupation of glorifying the Creator by observing the wonder of his works'. The king capitulated, and Masson, full of vim, set off for a country which he felt was now at peace. He was escaping all the wranglings within Europe as France was still struggling to find a stable government in the aftermath of the French Revolution. He did not anticipate, of course, encountering anything like the irksome restrictions imposed on him by the Dutch in South Africa. In fact, he expected plain sailing.

He left in September 1797, before the ship was within sight of the Americas he was being ricocheted between strange ports, with his high spirits decimated and his composure evaporated. He had convinced Banks and Aiton that with a lifetime's worth of wisdom behind him he could tackle just about anything, anywhere. He had been seized once before. But despite the propensity towards lawlessness on the high seas, and in spite of his own previous experience, he was aghast to find that he was once more seized against his will.

This time, the ship was halted by two French privateers, which left after boarding and examining their papers. Breathing a sigh of relief, they then made haste and proceeded without incident until 8 November. Then, three ships bore down on them, one of which was a French privateer hailing, they deduced, from the Caribbean island of St Domingo. Shots were fired, small-arms fire was pumped into the ship, the crew capitulated, and the pirates promptly boarded the vessel and captured the passengers and crew. All the passengers, including Masson, were bundled onto a conveniently passing vessel from Bremen bound for Baltimore.

For a gruelling and tedious three or four weeks the passengers were crowded together on board this ship, exposed to the elements and deprived of all but essential food and water. They were then transferred to yet another ship, and finally disembarked at New York. Even with a lifetime of narrow squeaks in foreign lands under his belt, Masson was clearly shaken by his ordeal, writing to William Aiton at Kew of 'his great distress after experiencing many difficuties'. But if he had any thoughts about giving up and returning home he remained silent on them. He set off straight away for Niagara. From there, Masson traversed immense areas of Canada, exploring the shores of Lake Ontario before returning to Niagara and from thence to Montreal. Again Sir Joseph Banks had loaded him up with letters of introduction. In the summer before Masson departed, Banks had taken the precaution of writing to the American Ambassador in London, Rufus King. He had prepared, on behalf of the king, an eloquent citation of Masson and his triumphs in the botanical world. He informed the ambassador, and anyone else whom it might have concerned, that Masson's remit was to explore the tract of high land between Lakes Michigan and Huron. Banks, ever the diplomat, emphasised that his eminent plant collector clearly understood the boundaries of the brand-new United States, but that he hoped that if, perchance, Masson inadvertently strayed over a line, purely of course in the interest of botany, this would be overlooked. Banks was clearly remembering Masson's waywardness in South Africa.

That time, Masson had got away with it, but this time it might be very different.

Banks was right to be concerned. Totally in keeping with his character, Masson was resolute that he would not be confined by the constraints of political unrest if he wished to explore widely. He plunged into this new territory, determined to make the most of his time. Previously he had been under pressure to work hard for his masters. On this occasion he felt able to take time to paint scenes such as the Niagara Falls, he botanised widely and, by the by, managed to collect or help collect a menagerie of animals. He did not venture widely out of his chosen area, and his natural curiosity now included fauna as well as botany.

Colonel George Landemann of the Royal Engineers was busy moving from Drummond Island, now in the possession of the Americans, to the island of St Joseph. He had been supplied with a flat-bottomed boat 36 feet long and about 7 feet wide and sharply pointed at both ends. On board he had a cargo of wild animals consisting of two bears, a deer, a blue fox, a beaver, a raccoon, an otter and various other small animals, all in good cages and properly secured, although puzzlingly, there is no record of any reason as to why he was transporting this collection. In his memoirs *The Adventures and Recollections of Col. Geo. T. Landemann*, which were published in London in 1832, Landemann also described that he also had some 'very fine logs of curled and birds eye maple'.

At some point, Colonel Landemann seems to have encountered Masson, although he seems to have been somewhat unsure as to who exactly this 'British naturalist' was. He assumed that he was employed in some way by the British government, and was about sixty years of age. He found him kind, very plain and unassuming and thought him very scientific. His account of Masson is untainted by any knowledge of his qualifications or background: 'He accepted of the shelter and what trifling nourishment I could provide, with unaffected diffidence, and submitted to the thousand inconveniences of the want of every comfort, without an expression of regret, excepting his frequent mention of the fear he entertained of incomoding me.'

Masson roamed over vast areas, for example the 'immense tract of high land between Lake Michigan and Huron', but, other than the above encounter, almost nothing is known about exactly where he went during his seven years there. Owing to the severity of the winters which buried most forms of plant life under snow, it is hardly surprising that he took immense interest in the animals surrounding him. However, he did have some botanical success and he is credited with sending back the nut of a pecan, with a note 'that it had already met with several accidents, and might need putting into a hothouse to force it'. He also sent back one outstanding plant, a species of *Trillium* (trinity flower or wood lily). As usual, Masson had found a very unusual species. Trinity flowers, or wood lilies, have flowers and leaves in threes. They have a double appeal because they carpet woodlands with sweet white flowers and because nature produces few plants with three petals or leaves. They have captured the imaginations of generations of gardeners seeking something a little different.

We do know for certain that on 30 December 1805, Francis Masson was alone in Montreal. He took to his bed and within a few days he was dead. His death provoked a flurry of activity amongst the necessary lawyers and curators appointed to look after the interests of his two nephews in Scotland. At first there was a petition by the Supreme Court in Montreal to bring forward an official to deal with Masson's death. In this petition he was described as being 'Father' Masson, given the title of a priest. There appeared also that same year a note of one Louis François Rodrique Masson, an employee of the fur trading North West Company, whose name was similar but of no connection. From this confusion might well have come the rumour that Masson was French or of Huguenot descent. Further documents extracted from the records of the Presbyterian congregation in the city of Montreal, from their register book of 1803–09, show that the Court appointed four eminent officials to be the curators of his 'estate'.

So having established in Montreal that he was neither French, nor a priest, his worldly possessions were sorted out, his savings left to nephews Francis Alexander, the son of his sister living in Kemnay, Aberdeenshire, and John, the son of his brother Charles Masson. The location of his grave is unknown.

This lonely death on foreign soil seems to have provoked some reaction at home. His old friend James Lee from the Vineyard nurseries in Hammersmith voiced the accusation that his health must have suffered grievously from exposure to the bitter temperatures of Canada, after a lifetime accustomed to hot climates. And all, as Lee remarked with some bitterness, 'for a pittance'. James Lee's son wrote to Sir James Edward Smith, the president and founder of the Linnean Society, explaining that 'What Masson has done for botany and science deserves to have some lasting memorial given to his extreme modesty, good temper, generosity and usefulness. We hope when the opportunity comes you will be his champion.' Three months later he wrote again, but perhaps Masson had been out of sight for too long, and there were other young adventurers eager to step into his shoes.

There is, however, a lasting memorial to his life. He may have died in the freezing winter conditions of Canada, but his legacy is vibrant with brilliance, scent and sunshine. He brought us blue water lilies, red hot pokers, the drama of amaryllis and the evening scents of gardenias, along with many heathers and curious *Massonia*. His *Gladioli* and the dry, papery, everlasting flowers of the *Xerantheum* and *Gnaphalium*, both of which bear the evocative title of 'immortelles', have transformed planting everywhere, from humble cottage gardens to grand palaces.

He also solved mysteries. 'Naked ladies', the *Nerine sarniensis* or Guernsey lily, had been so named because it had been washed up on the shores of Guernsey in the Channel Islands. Masson found the original in South Africa.

Ironically, the proteas that he so zealously collected and cherished became the national flower of South Africa. Pelargoniums still sell in their billions each year whilst his *Zantedeschia aethiopica*, or arum lilies, with their white trumpet-like flowers have become wedding flowers for the fashionable as well as a must-have for the water garden. And the curious cycad, the oldest pot plant in the world, which arrived back at Kew in 1775, has finally fruited, although only once so far. Its cone was viewed by Sir Joseph Banks when he paid his very last visit to Kew as an old man just before his death in 1819. It is anyone's guess when it will bear fruit again.

CHAPTER 4

William Wright (1735–1810)

BRINGING THE WEST INDIES TO EDINBURGH'S WEST END

———

If Masson was the gatherer of sunshine flowers from a glowing climate, William Wright, a doctor from Crieff, was to lavish time and brain power on the cure of tropical diseases. As Masson sailed off on his traumatic last trip skimming the West Indies, Wright was breathing a deep sigh of relief; he was finally on his way home from Jamaica. As Dr William Wright had boarded the *Barton* in Barbados bound for Liverpool he was demob happy and euphoric. At the age of forty-nine he was on his way home for good, with some financial security and a deep longing to see his native Perthshire again. Despite spending most of his life at sea or latterly as a doctor on the island of Jamaica, he had remained ever homesick.

In contrast to Masson, who constantly pined for more adventure, Wright longed for nothing so much as to settle back into his small Scottish home town, and be fussed over by his chattering nieces, the fashion leaders in the small burgh of Crieff. As he boarded the ship in April 1798, he anticipated nothing untoward. His life, at last, looked clear-cut. As he looked east to the horizon, he could almost smell the moist, mossy soil of his native Perthshire. But like Masson, who had passed by a few weeks before, he was in for a rude awakening. The French frigate *Le Tigre* was loitering around the coast of Ireland – possibly the very same frigate that had briefly shadowed the ship on which Masson was sailing westwards. Whilst Masson's ship was later boarded, Wright was luckier; the *Barton* made a run for it, and escaped. So Masson, the man whose life had been spent in hot countries, and Wright, the doctor whose life had been spent, not of his own choosing, in the heat of the West Indies, were aboard ships that almost literally passed in the night. They had most probably heard of one another, for yet again Sir Joseph Banks was a common link. But Wright was a very different character to Masson, and came from a very different background.

Born in Crieff in March 1735, he was the son of James Wright, a blacksmith who had earned passing fame as the man who had shod the horse of Charles Edward Stuart, Bonnie Prince Charlie, as he passed through Crieff in 1746. When his infant son William was a year old, James, a man of some independent thinking, warmth and determination, had completed the erection of his new house, and with pride inserted a stone lintel above the door with the emblems of his trade. These were a hammer, a horseshoe and a pair of pincers. The stone also bore the date, 1736, the initials I.W. and, for good measure, a thistle and a crown. The street bore the name King Street. By 1898 the hand of the local authorities had descended on the building, demolishing the original, and erecting a police station with 'light airy cells' for the prisoners, situated conveniently adjacent to the spacious courtroom. Anxious to reflect the growing status of the town and, in the words of Andrew Porteous, a contemporary author who has written extensively on Crieff, to uphold the 'dignity of the ever increasing burgh', James Wright's lintel stone was saved and inserted over a doorway. Within was a portrait of William Wright who was by then a renowned figure in Crieff, in Scotland and in the world of medicine.

James the blacksmith must have done well, despite backing the Jacobites. He survived the sacking of Crieff, after Bonnie Prince Charlie held a council of war in the Drummond Arms Hotel in 1746 before retreating westwards. James married twice, and his sons from both marriages were well educated. Scotland's oldest public library, at nearby Innerpeffray, which had been founded in 1691, was well used by the inhabitants of Crieff. William was the younger of two sons from the second marriage, and he and his brother James were bright little boys. With more than a nod towards their father's political interests, and possibly a well-honed ability to eavesdrop on adult conversation, they sneaked into the local history books aged ten and twelve by meriting a mention in the Duke of Cumberland's dispatches.

William and James, to whom William remained close all his life, formed themselves into a two-man mini-army and let off their toy guns just as the staff and flag were passing over the bridge entering the town. Fortunately, Cumberland was heard to remark that however the adults might regard this arrival of the English army, at least this little episode appeared more of a welcome from the children. Both boys and their family seemed none the worse for the misadventure.

Crieff in the 1740s was a thriving market town, one of the 'trysting' places where the black cattle were driven south of the Highland line to be sold on to English buyers, who were justifiably fearful of venturing into what they imagined to be the lawless Highlands. James, two years senior, was happy to remain in Crieff all his days, but William saw his future elsewhere. At the tender age of seventeen he was apprenticed to a surgeon in Falkirk, and by twenty-two he had arrived in Edinburgh to study medicine, staying at the house of his uncle, from where he also managed to wangle a trip to Greenland at his uncle's request. As yet, there was little sign of the homesickness that was to plague him all his life.

Edinburgh in the late 1750s was home to an organised medical teaching school of thirty years' standing. Among the lectures were those on anatomy and Wright had to steel himself for the unpleasant aspects of dissection and surgery, especially if examples were available to work on. Alexander Munro, the Professor of Anatomy in Edinburgh in 1725, vehemently had abhorred the 'vile, abominable and most inhumane Crime of stealing human Bodies out of their graves . . . And do, for encouragement to the Discoverers of such Violators of Sepulchres, and of other malicious Felons, who endeavour to bring a Reproach on my self, etc etc . . . and hereby Oblige myself to pay a Reward of Three Pounds Stirling for every such offence that shall be discovered to the Magistrates, so as the Offenders, one or more may be convicted.' In 1752, Helen Torrence and Jean Waldie were hanged in the Grassmarket for murdering a young boy in order to sell his body to medical students.

Ten years earlier, Dr William Hunter had studied in Edinburgh and been dismayed to find that:

> In the course of [his] studies, [he] attended, as diligently as the generality of students do, one of the most reputable courses of anatomy in Europe [that of Munro above]; there I learnt a good deal by my ears, but almost nothing by my eyes; and therefore hardly anything to the purpose. The defect was that the professor was obliged to demonstrate all the parts of the body, except the bones, nerves and vessels upon one dead body. There was a foetus for the nerves and blood vessels; and the operations of surgery were explained, to very little purpose indeed, upon a dog.

If Wright had to put up with little more than this in the way of anatomy, he also had to survive three years' study which included chemistry, botany, *materia medica* and pharmacy, surgery and the theory and practice of medicine. Just as challenging was the social aspect of student life. Wright was among no more than 250 students studying medicine at Edinburgh. Indeed, as his father was a mere artisan, a blacksmith, he was immediately at a disadvantage. For the sons of artisans, studying medicine was unusual. Medicine was mainly for the offspring of the middle classes, such as ministers, lawyers, military men and, of course, physicians themselves. For the 'lad o' pairts' (talented youth) whose intelligence and academic potential shone out at the local village school, but who came from a very ordinary, lowly background, the Church was the traditional route, and the path there was aided by financial bursaries. Medicine did not have the same cachet. That said, a few lads of poor background did manage to study medicine as a first degree. Often they came to it via previous study – often divinity – and could support themselves through study by tutoring. Wright was lucky to be able to board with his uncle.

So, without independent financial means, though even with a bona fide medical qualification, establishing oneself in the medical world was far from straightforward. As Wright was quickly to find out, the money spent on his student qualifications left precious little to start up a practice, and an Edinburgh degree was simply not good enough for the British Army, who favoured degrees from Oxford and Cambridge. Then there was the cash required to buy a commission in the army, and, generally, the private income required to sustain an army officer. Entrance into the British Navy was not easy, but for a man of no independent financial means, it remained one of the few opportunities open to a newly qualified doctor, and a tried and tested route for a young Scot with his medical degree.

Towards the end of his studies in Edinburgh, Wright penned a letter to his father thanking him for his liberal education, and in a will, detailing that he was assigning his legacy from his parents on their death, straight to his brother if he, Wright, predeceased them. He was grateful for his support so far, but he was confident he was now well on the road to independence. In a very different vein, however, he confessed to his brother James that he felt nothing like so sure of his future. Goodness knows if he would succeed in passing his naval entrance exams or indeed get a post at all, but it was best not to alarm his parents. It was all an adventure, and he was off to explore the world.

With his graduation promise still ringing in his ears, having sworn 'not to wantonly try experiments with any of our patients, nor yet divulge any thing that has been told to us as a secret, but exercise all the duties of our profession "*casti et probe*" – honestly and virtuously', he left Edinburgh to seek his fortune in a gang of like-minded recent medical graduates. In high spirits they sailed from the port of Leith, close to Edinburgh, huddled in the cheapest of berths, and were dumped unceremoniously on the Essex coast, from where they had to find their way to London.

Again, Wright at least had a relative to put him up. His elder half-brother gave him board and lodging while he waited to be examined at the Surgeons' Hall. A fever nearly put paid to his chances of getting there, and his half-brother's household made no bones of the fact that they felt imposed upon, rather than happily helping a relative to better himself. Weakened by illness, Wright still managed to attend his oral examination.

He was examined by a naval surgeon who had sailed down from Leith on the same ship, and was marking time in London by examining students while awaiting a better appointment. Wright passed, succeeding in detailing the thorax, but, in his own words, forgetting the 'covering of the heart'. But there were a few nervous days' wait before he was summoned to the Navy Office, and appointed at the going rate of fifty shillings a month as a second mate on the Royal Navy ship *Intrepid*. Lord Breadalbane, who owned a vast tract of land in Perthshire, had also put in a word for him, and with this he clinched his post as a junior surgeon, a more elevated post than second mate.

Things got off to an inauspicious start as, before the ship had left the dock, it was Wright himself who was ill with 'jail fever', a common enough umbrella description of various ailments, such as a high fever, cough and diarrhoea. His illness was hardly surprising as he reported that the *Intrepid* was full of 'the refuse of mankind and the very dregs of the human race' confined within the 'plaguey bulk containing the sickliest of the Royal Navy'. He had to be taken ashore to the hospital to recover. 'Oh! My brother,' he moaned to brother James in Crieff, 'never come to this wicked place but settle in your own country, so you may live happily. Had I been made some mean mechanic, I should not have had the occasion to range the world in search of bread.' On the upside, the *Intrepid* had a good captain, Pratten, who had been given the temporary rank of commodore, and an amiable Irish surgeon, Pierce Butler, who was 'good natured and well bred in the extreme'. A student friend of Wright's, George Eason of Dysart, was also on board.

The *Intrepid*, along with a squadron of six ships and a couple of frigates, rattled round the Channel and then the Mediterranean. There were bonuses. Britain and France were in the midst of the Seven Years War and on 26 April 1758 off the Isle of Rhe near Gibraltar, the *Intrepid* captured the *Raisonnable* producing some useful prize money. Then there was victory over the French fleet under de la Clue the following year off Cape Lagos. Wright also began to evolve into a travel writer, describing the Egyptians, and Armenians and their habits and costumes, and waxing lyrical about the shores of the Mediterranean.

In these heady and early days, the young surgeon William Wright honed a useful skill. He found he could very easily make friends and influence people. He might only have been the son of a blacksmith, but he possessed a slim, patrician face with a long Grecian nose, and was quite unlike the ruddy-faced Francis Masson or the stockily built William Forsyth. He was ambitious and observant. So he made sure that he kept up his journal of life on board the ship, closely observing the cleanliness habits of the crew, and how thy kept food fresh on board, and storing these details away for future use. This was common practice on naval ships, and ambitious young men would use these observations as an extra learning tool. In his early twenties, he took stock of his place in the social hierarchy of life and determined to climb high. He fully realised that the way forward in life was not purely through education: friends in the right places higher up the social scale were the key. A social climber he undoubtedly was, but to his credit this was tempered by his genuine liking for most of his fellow men. Most of his contacts and acquaintances did remain genuinely lifelong friends. If they could also open a door for him, through which he saw a better future, so be it.

Supported and encouraged by his immediate superior and by a surgeon named Butler, now also his friend, he applied himself to renewed study, and passed enough exams to gain an appointment as a first mate on the *Danae,* a forty-gun frigate under Sir Henry Martin. Ebulliently he wrote to his brother that the ship was not only the finest in the navy at that time, but that he would 'soon surmount all [his] difficulties and . . . make you all [the family at Crieff] comfortable'. On board the *Danae* they chased up to Scotland, seized privateers 'infesting the Western isles', fell in love with the people of Ireland and throve under the benevolence of the captain, Vaughan, who liked nothing better than to have all his officers round his laden table for gourmet dining. Wright felt as though life was proceeding smoothly. On the brink of finally reaching the peak of his medical qualifications, he felt confident that his promotion to the rank of naval surgeon was right round the next corner. However, while in this relaxed atmosphere at Cork, the ship received orders to proceed to the West Indies, with Wright still a junior surgeon. The longed-for promotion proved elusive. But if he wanted to see more of the world, his opportunity was only an Atlantic's width away.

Wright arrived in the West Indies in time to see the scuttling of French hopes to secure the islands for their own. In quick succession he witnessed the fall of Martinique, then Grenada, followed by the tumbling of St Vincent and St Lucia. He shuttled between the *Danae* and shore-based hospitals at Fort Royal and St Pierre, then on to HMS *Culloden*, and the frigate *Levant* whilst island-hopping every couple of weeks. All the while he was scornfully noting the outdated French medical methods. For himself, his own medical experience was widening by the day, while his hopes of promotion continued to wane. Although he had been gathering friends inexorably, from Doctors Saunders and George Munro to Gartshore, and came under the wing of Commodore Sir James Douglas, who rapidly became 'my friend Sir James Douglas', he somehow never seemed to be near the right ship at the right time in order to gain that leg up the ranks to surgeon. His frustration surfaced rarely, but he wrote tersely about his senior surgeon on the *Levant,* who although an amiable character of many talents spent much of his time in an alcoholic stupor. For Wright it was work and more responsibility with no increase in pay or advancement.

Within a couple of years he had returned to Britain. The Seven Years War was over, young medical officers such as he were two a penny and he found himself unemployed. Nevertheless, he pressed on. He had not even the resources to travel to Crieff to visit his parents. It was a bitter blow to his ambitions to advance in the navy and establish his own medical practice in Scotland, but he had long realised that by studying when waiting for his fortunes to change, he might stand a better chance of succeeding in life. In the meantime, he moved into the London home of a Dumfries man, Mr William Collart, also a surgeon from the *Levant*, but 'a good man', unlike his drunken superior. Together they planned a medical partnership somewhere in North America, but the colonies of North America were seething with unrest in 1764, and for a pair of young doctors still attached, if only on token pay, to His Majesty's Royal Navy, it was not an auspicious time to emigrate.

Reluctantly, the 29-year-old Wright decided to sail for Jamaica and seek his fortune. Collart generously donated a valuable medicine chest to help him on his way. Just before sailing he had been challenged about his bachelorhood by his widowed uncle but Wright was ready with the chirpy response, 'Make my compliments to your intended!' To his brother he jokes, 'well then, when all my sweethearts have forsaken me, what say you to my attacking some rich widow, and making my fortune by a *coup de main*?' The truth was that he had not even enough money to rent a property, let alone consider marriage.

Almost on cue, the voyage to Jamaica on board the *Bonella*, commanded by yet another friend, Captain Duthie, was enhanced by three lady passengers. Calling in at Madeira for a couple of weeks, the party toured the island. Wright was enchanted to find that much of the flora and history were well documented in a recent book, enabling him to search out the most interesting sights and plants. Wright was in his element; plants and female company suited him. Fortified by a pleasant three-month cruise, his arrival in Jamaica in March 1764 was a deflation. Jobs for surgeons were few and far between, and Jamaica was overflowing with doctors just as well qualified as he, who were reduced to serving as assistants at around £40 a year, especially in the Savanna le Mar area, which had been his first choice.

But he immediately busied himself knocking on doors and chatting to likely employers, and his prowess at making successful contacts paid off handsomely. Firstly he found a job as the assistant to the principal surgeon in Kingston, Dr Gray, at £100 per year, plus board and lodging, which he was quick to point out was expensive on the island. When this first six-month contract was nearly over, he accepted an offer to become a partner in a medical practice from another old acquaintance, Dr Tom Steel, whose house was on Hampden Estate in the parish of St James', about a hundred and fifty miles from Kingston. By this time Hampden was owned by Mr James Stirling and managed by Patrick Stirling of Kippendavie, a family estate adjacent to Dunblane, and fifteen miles south of Wright's home town of Crieff.

Jamaica of the mid to late eighteenth century had a reputation as a seedy, whore-ridden, dangerous island, with Kingston vying with other ports for being the most dangerous in the Caribbean. Sugar production was labour-intensive and required large populations of slaves. Although Jamaica lagged behind Barbados when Wright arrived, this was about to change, and Wright arrived on the cusp of expansion. Within twenty years Jamaica was to become the greatest sugar exporter in the world. The partnership was a shrewd move for Wright, as the income from the practice ballooned owing to the increasing numbers of slaves in the immediate vicinity; for each slave on their register they received a set fee per year, whether the slave required their medical services or not. For each of the 1,200 slaves on the surrounding plantations, they received 5s. Added to this was a practice covering all non-slaves living within a twelve-mile radius. Their income assured, they felt able to invest for the future. Within six months they had bought £500 worth of household furniture, seven horses and four slaves. They added on another hundred acres to the twenty around the house, paying out £391 and giving themselves a property of close to fifty acres.

With the security of a regular income, Wright at last felt he could indulge his interests in natural science, and receive some reward, if not financial, then as a valued contributor to the natural science collections at Edinburgh University. His offers of sending examples of birds and insects were received with alacrity. On the medical front, his star was also rising.

The outbreak of smallpox in 1768 gave him a chance to try out a treatment first practised on the Guinean coasts, in which the body was covered with wet clay to try and bring down its temperature. Wright expanded this theory of cooling fevers by throwing open doors, letting in fresh air, and dipping the patient in cool water every few hours. A couple of years later, in March 1770, he owed his life to both the success of this method, and the willingness of his medical partner Tom Steel to apply it. He recovered, but in the aftermath of this illness he longed to return home. 'We are both heartily sick of this way of life, and long for the time when we can leave it with a good grace, that is, when we can do without it,' he wrote. But this did not prevent the partners from building another house, Orange Hill, in 1771, and hiring a white man to oversee their slaves, now numbering thirty-three.

Dr Johnson was railing on one side of the Atlantic that Jamaica 'was a place of great wealth, a den of tyrants and a dungeon of slaves', but Wright allowed none of this controversy to seep into his correspondence. Instead he concentrated on his herbarium, which eventually numbered 761 specimens, some of which are still in the possession of the Royal Botanic Garden in Edinburgh.

The following year he recorded a detailed account of his methods of reducing fever. Called to the bedside of thirty-year-old cooper William Jewell, Wright found him in a room with curtains and windows tight shut, 'stewing with warm drinks and under a load of bed cloaths [sic]. His headache was great, his thirst intolerable, his skin burning hot', but by gradually removing the bedclothes, opening the Venetian shutters and giving him sips of water, his fever subsided and within a day he had totally recovered. By 1774 Wright's medical expertise was rewarded by the appointment – one of honour and distinction, but purely honorary as he promptly pointed out – as Surgeon-General of Jamaica.

Parallel to this honour was his introduction to the new governor of the island, Sir Basil Keith, and the following year, in 1775, he discovered on the island the *Cinchona jamaicensis,* a species of the Jesuit's bark, the inner bark of which he recommended as of equal effect in the treatment of fever, today accurately known as malaria.

Later in 1775, his paper on the treatment of diabetes was published and read before the Philosophical Society of Philadelphia. Diabetes baffled contemporary doctors, and their efforts at treatment probably killed many of their patients, varying as it did from one to all of the following: opium, blood-letting, emetics to induce nausea, and a type of diet consisting of large quantities of animal proteins. Wright's remedy consisted of lime juice saturated with sea salt. It may have helped some conditions, but it certainly had no long-term effect on diabetes. It was much admired, though, by the medical men of the day and its complete ineffectiveness only came to light long after Wright had retired.

By now it was nineteen years since he had been in Crieff, fourteen years since he entered into his partnership with Tom Steel in the West Indies, and a couple of years since Steel's marriage to the daughter of a neighbouring plantation owner, since when an innocent *ménage à trois* had apparently little altered their professional and personal harmony.

Every year Wright repeated his promise to the family to return home, but his timing was, as ever, awry. His fortunes were at the mercy of the American War of Independence, which had been raging over the previous couple of years and had resulted in a drastic fall in the rate of exchange. His years of investment in Jamaica, his house, land and slaves simply would not realise an adequate enough amount for the longed for return to Scotland. Finally, by July 1777 he could wait no longer. Cutting his losses, he set sail on 1 August from Montego Bay bound for Liverpool on board the *Thomas Hall*, amid a fleet of seventy-six merchant ships, protected by three warships. He left behind the bulk of his investments, taking only enough cash to tide him over for a few months. Three weeks into the voyage, a storm, his worries about his financial future – resulting from the unremittingly bad news drifting in about the War of Independence in North America, which was severely eroding his savings and investments – and his exposure to a dramatic drop in temperature as the ship left the tropics behind, rendered him seriously ill.

By the time the ship docked in October he had no energy to proceed straight to Scotland. Instead he went to London. As usual, though, he found a friend with whom to lodge, and recovered from his ill health. He settled in with another long-established associate, Dr Gartshore, who ran a lucrative obstetrical practice in St Martin's Lane, and rekindled his association with Sir Joseph Banks and Dr Solander. With such associations, he slipped with ease into the higher elements of society, who invited him to present a talk on the cabbage bark tree of Jamaica and show off his collections of plants which he had brought with him. His 'talents for conversation which enabled him to make his knowledge available at all times' stood him in excellent stead. He was admitted at lightning speed to the Royal Society of London and passed satisfying hours at the Royal Botanic Gardens at Kew. He was able to watch the development of the plants he had previously sent over to William Aiton which were now growing in the greenhouses, and happy times were passed imparting advice on the climatic and soil conditions from which the plants had originated.

Wright had a good turn of phrase, a light-hearted approach to much of life, a happy disposition – when he wasn't miserably homesick – and was clearly an amusing companion. Years later, he carefully mounted his herbarium collection, much of which dates from his first sortie in Jamaica. The rough paper with unevenly edged tissue paper separating the pages is beautifully laid out. Some two centuries later, the pages still turn easily, although the occasional specimen, such as number 135, *Pimenta*, the Pimento or Jamaica pepper, have out-stepped their boundaries. The pepper's seeds have spilled out of the pods and elegantly eaten through a couple of pages, leaving the centre like a lace doily made from a brown paper bag. The seeds, still intact, have slid into the central binding into the shiny hoof glue which still holds the entire book intact.

Wright was an educated man, but his early life as a small-town Scot of artisan origins and his many years as a plantation doctor had clearly taught him to connect easily with folk. The book might have been catalogued according to the Linnaeus system, and contained erudite Latin terminology, but his descriptive text shows his mastery of the common touch:

Number 141, Mamea or Mamee Tree
(Mammea americana L (Guttiferae))

The fruit or apple as large as a man's head of a gray or brown colour, the inner rind . . . [unintelligible] . . . looks and tastes like a carrot. And is much liked by people in this country – wild hogs feed on it. Careful way wardens never suffer this tree to grow near the roadside, for should this ponderous fruit fall on man or beast it would break their bones and kill them on the spot.

Wood is good building timber which stands wet weather better than any other and used here as cisterns for molasses.

Or:

Number 83, Alligator or Poosoo Bark Tree (Daphne lagetto)

This tree grows on rocky hills – almost inaccessible blossoms small and numerous.

The bark of the tree was long known to the rebellious negroes under Colonel Cudjae, before their capitulation in 1739, they still procure it and sell it to the white people.

A straight piece of the trunk being cut to a proper length is beaten with a smooth stick all round. The bark is then pulled off the outer stem, separated as useless, and the rest put in a pail of

clean water, where it is soaked for a few hours; and rinsed with
fresh water. Before it is quite dry, begin to separate the Laminae
or layers from one another . . . then get clean gauze.

Thus has Dame Nature furnished a cloth, ready woven and
bleached. Our ladies make it into Caps, Ruffles and even suits of
this curious bark. If managed carefully it will bear washing with
soap and water.

In medical circles, his theory, proven practice and success rates
of reducing fevers in the tropics by cooling the patients also
provoked much interest. But conservative prejudice was strong in
the established medical world, and he had to wait many more years
before this theory was widely published and accepted.

Life would all have been satisfactory if Wright had been able to
put a fortune in his back pockets in Jamaica and return to Britain
to enjoy a life of leisure, giving learned talks here and there and
enjoying the society of men of letters. But he needed to earn a
living, and try as he might, London was well staffed with medical
men of mainly tropical experience. He had many friends, but no
job offers. He hesitated, turning over the possibilities. Finally, as
London could not offer him work, and after much persuasion from
his family, he decided to put his pride in his emptying pocket. In
January 1778, he left for Scotland.

His homecoming was to be very different from that which he
had imagined for all those years in the steamy heat of Jamaica. His
brother James, with whom he had remained close through twenty
years of separation, had built a house for his return. Amongst the
Crieff family there was an undisguised pride in having a sibling
who was apparently successful and hobnobbing with some of
the great men of the era. His house was already christened 'Dr
Wright's House'. Brother James and his family had moved in to
prepare for the returning wanderer. However, Wright had returned
with neither the savings to assist his parents in old age, nor any
confidence that he could afford to maintain his new home. He
managed to turn the tables, and insisted that his brother's family
remain in 'his' new house, and he, William, would be their lodger.

The family ties were strong, the new arrangements worked well and Wright spent time teaching his nephew James, and contributing financially to his education from his meagre resources. However, all the while, he fretted about the situation in the West Indies and the war in North America, and it was difficult to hide his anxieties. He wondered if he would ever retrieve a reward for his twenty years of hard work and investment in Jamaica. After a year in Perthshire he moved to Edinburgh, still worrying. He presumably must have found some work in Edinburgh, but it appears to have been insignificant.

At home in Crieff, his brother James was concerned about his son James junior, and pressed his brother to offer him advice on a career. Wright sent the following touching reply: 'I shall be glad to hear from Jemmy. His profession in life must be left to himself. I wish it may be one that will not oblige him to wander.' It was just such wandering which Wright himself could not escape. On 18 September 1779 he writes that a squadron of French men-of-war had been cruising in the Firth of Forth for several days, swooping and capturing locals, and threatening to land in nearby Leith. On 23 September he confirms the safe arrival from brother James of a requested 'sword in good condition. He that would not draw it in defence of his country is unworthy to live in it.'

As there appeared few other suitable prospects for a doctor with expertise in the tropics, little income and a penchant for plant collecting, Sir Joseph Banks encouraged him to join up again. Uneasily he found himself drafted into a new corps named the Jamaica Regiment, although he was reassured that it would only be obliged to defend the island. Wright was appointed the regimental surgeon. As usual, he was assured by Lieutenant-Colonel Balfour, as they proceeded from London to Warwick, that it was a fine body of men, but on arrival he noted with sickening clarity that his fine new sailors had been conveniently extracted from London's overflowing prisons. Wright even recognised one of the mutineers who had sided with the French the previous year at Edinburgh.

Under the command of General Monkton, a flotilla comprising

fifty-five unarmed merchant ships and various frigates and warships set sail in convoy. Two days later a couple of the naval ships left the main group, and by the time the fleet entered a dense fogbank off Cape St Vincent, Wright's ship, the *Morant*, found herself almost alone facing the substantial combined French and Spanish fleet. After a small skirmish in which, curiously, the only casualty was the regimental commander's wife, the ship was captured. Wright and other officers were permitted to take with them their personal luggage. Wright wrapped up the regimental flags undetected and when they were transferred to the French *Bourgogne*, managed to destroy this evidence. By so destroying the flag, the ship was therefore no longer a strategic, naval, political and much valued prize, but merely a vessel worth its salvage value.

Eventually the British prisoners were released from their slovenly, 'undisciplined' French captors, whose evening rantings about the supremacy of the allied French and Spanish fleets grated on the captive audience. As the officers were transferred into the custodianship of the Spanish, their French captors demonstrated a somewhat embarrassing late rush of fondness for the British captives. Wright, with some embarrassed wriggling, extracted himself from the enthusiastic Gallic farewell embraces of the man in charge, Captain Marien, and he and the rest of the officers were marched by their new Spanish captors inland from Cadiz, where they had landed, to Arcos de la Frontera in the province of Andalusia. On the way, they were singularly unamused and even displeased to reach the town of Jerez de la Frontera, too late in the day to persuade their host to provide any of the town's famous 'white wine'. They had to content themselves with sherry, which they quickly judged far too young for drinking, but managed to swig it down accompanied by a meagre supper of bread and cheese, with a seasoning of garlic. Prisoners they might well have been, but they drew the line at slumming it.

In the following few weeks, they passed the time criticising the laziness of the locals, whose work ethic, they concluded, was negligible. They clamped their handkerchiefs over their offended noses when jostled by the masses – presumably Spanish, although Wright does not say – who had been expelled by the Moors, and who thronged the streets. The olive presses, Wright observed, were inferior to the linseed presses in Scotland, but the sheep throve due to excellent irrigation.

As time wore on they were allowed to ramble around the countryside within a six-mile radius; they fished for eels and mullet, and shot some of the abundance of hares, rabbits and birds, all this while reserving their chief energies for hunting wolves – although Wright does not record if they shot any. It all sounded like a happy country-house hunting party.

For his part, Wright searched the riverbanks for replacements for his herbarium which had been a casualty of the capture, and practised his medical skills on a very grateful local populace. Medical practices were a century behind those in Britain and he was appalled to find blood-letting and warm water were the treatments for most illnesses, especially fever. The late summer heat allowed him to make useful comparisons between the local illnesses and those he had seen at close range in the tropics.

Eventually the prisoners were repatriated, and Wright and the regiment embarked once more for Jamaica, where, to his relief, the political situation had calmed down, and the mangy throng of the regiment, many of whom had escaped capture and had been stationed in the West Indies, was disbanded and returned to Britain. Wright managed to remain behind in Jamaica and made his way to a warm welcome at Orange Hill.

Life picked up as before, but within a few months his old friend Tom Steel suddenly died from a fever which even Wright's expertise with cold-water treatments could not salve, and he found himself as executor on behalf of Mrs Steel and their five children. He managed to speed up the whole process, using his local good name to good effect and in record time he settled not only the affairs of the widow and her children satisfactorily, but also his stake in the investments within the property and medical practice. Aged forty-nine he was in reasonable health, and with a healthy bank balance. It was now 1784, the American War of Independence was over, and investments in the Americas and the West Indies were now, owing to the stability of the political situation and ease of trade, of much greater value in Britain. Property in the West Indies was also of value, as investors in Europe sought once more to venture across the Atlantic.

With free time on his hands, Wright embarked on his much delayed task of gathering together again a duplicate of his original herbarium, detailing every single plant on the island. He also cast his eye over the introduced plants, and eventually toured nearby islands to enlarge his collection. Together with Olaf Schwartz, a Swedish botanist soon to be famous, Wright worked away cataloguing and completing his herbarium. He was appointed as Physician-General on Jamaica, a post he held until 1785. Much of his contribution to botany was bound up with medicine.

Cinchona jamaicensis had been brought back by Sir Hans Sloane years before. By grounding up the bark, Wright had applied it as eye ointment. No one had deduced as yet, and were not to do so for many years, the real association between quinine, extracted from the bark, and 'night fever' or malaria. Although the application was wrong, Sloane's model of tropical plants being useful in medicine appealed to Wright. Busily he applied himself to other 'plants for a purpose' but was cautious in claiming anything a miracle cure. Wright was adept, when recommending herbal cures, in offering suggestions based on his experience, rather than pronouncing and claiming outright successes, and he had a gift for penning a short sharp résumé in laymen's terms: '*Capsia* [possibly *Cassia alata*] French Guava or Ringworm bush,' he scribbled elegantly, 'grows to ten feet. The flowers beaten and applied will remove the beginning of ringworm, but when inveterate we must look for other applications.' Still there amid the leaves of his herbarium collection within the library at the Royal Botanic Garden in Edinburgh, are the specimens and Wright's bold writing explaining the many deductions concerning his botanical finds and their application in medical matters, such as:

Alsine or Iresine the Bitter bush or bitter weed.

The leaves very bitter and antiseptic and of use in several weaknesses and in venereal diseases, but when disease hidden and yet advanced, maybe a turpentine and cold bath, will effect a cure; though at times it fails.

Lignum vitae (Guiacum officinale).

Decoctions of this wood are deservedly in repute [he writes approvingly] as it is also the gum for the cure of venereal complaints as well as [a cure for] the consequence of yaws and is also [a cure for] foul cutaneous [skin] eruption. The venereal disease is very prevalent amongst the negroes . . . but rarely found out – as it is well concealed – [unless one lays hold of the] spongy bones of the nose palate.

Myrtus pimento or Jamaica Pepper.

The ripe berries are black and larger than a currant and taste pleasantly aromatic but sometimes give violent headaches and I have known the same happen to people of delicate nerves in houses where dried pimentos are put up.

He finally sailed home, arriving at Bristol on 23 September 1785, bringing with him his vast collection of plants, settling with friends in Hampshire and London for the winter, busying himself with contributions from his collections to Kew and Sir Joseph Banks, and still finding time to oversee his nephew's medical studies. He also sent off his scientific papers on request to the American Philosophical Society, toured Scotland each year to botanise and accepted honours such as a Fellowship of the Royal Society of Edinburgh. He organised a sea passage for his nephew to accompany Stanley, a friend of Sir Joseph Banks, on an expedition to Iceland, and generously donated a copy of *The profitable Arte of Gardening*, 'Englished by Thomas Hill, Londoner, imprinted anno Domini 1574' to Banks in return for some botanical specimens from his Iceland voyage of many years previous. His interests ranged from discussions with Sir Joseph about engravings from a Dr Woodville, and various species of trees, to commenting on the French Revolution. A Roman Catholic priest, formerly at Dummond Castle by Crieff, and his niece had been executed in France, and Wright opined that the peasants there were 'satiating themselves with blood, and s[e]ize on riches which they have neither the talent nor industry to acquire for themselves'.

At one point, there was the possibility of a return to duties in the West Indies. The dreadful mortality of troops stationed there was becoming an object of national concern and Wright's name had been put forward as a person well acquainted with the area and its diseases. The Army Medical Board were hamstrung though by the Court Physician, Sir Lucas Pepys: Wright could only be appointed if he had a licence from the London College of Physicians, who would take several months to hand out their licence, and the fleet would certainly have already sailed. This provoked a heated correspondence between a Dr Wells and Lord Kenyon, and Dr Wright was duly appointed, setting sail from Portsmouth on 15 November 1795. Less than two days later, the fleet was decimated by a storm with the loss of 600 lives, but Wright survived, and this time was based in Barbados.

During his time on Barbados, Wright once again collected more plants and brought them home to Edinburgh, having by now one of the greatest private collections in Britain. Specimens included:

Mangrove Tree, Rhizophora mangle

[Lengthy description of this curious tree, with its limbs growing straight downwards into the water]. On this account there is no walking among mangroves, besides the dangers in some places of meeting with a hungry alligator who probably would make a meal of the Lord of the Creation and when once they taste human blood, are ever afterwards very ravenous. They are not now numerous because the country is more inhabited. Many amphibious creatures are destroyed and the vultures watch for their eggs and dig them up.

and:

Poke weed or Spanish Caliloo (Phytolacca)

Berries of the size and colour of a black cherry containing many seeds, and a rich purple juice. Many attempts to 'fix' this fine colour, but as yet attended with no success. Young leaves served up as greens, and preferred by some to spinach.

Food plants were also numerous: *Spondias purpurea*, the Brazilian or Spanish 'plumb', were stewed with sugar into a 'kind of mamalade [*sic*]' although Wright also thought that eating them with milk made 'an agreeable repast'. The *Laurus persea* or alligator tree, he noted,

> seems not native as never grows wild in woods . . . Produces fruit four times the size of the common pear and of the same shape. After removing the skin, one comes to a butyraceous [buttery] substance of a yellow colour. Eaten with pepper and salt and a bit of bread, tastes like marrow or butter but more palatable – few Europeans like it at first but soon become fond of it.
>
> Every person of whatever complexion cultivates this and no wonder since in the season it constitutes so agreeable and a principle part of their diet. Even horses, cattle, dogs, cats, rats, lizards, poultry and insects feed greedily of it.

The common name of this mysterious foodstuff? The avocado pear.

Cashew nut trees were acknowledged to be 'very pretty in bloom. From the Tree flows a clear gum like gum Arabic in taste and virtue. Some of the Barbadian beauties apply this to remove freckles, but in delicate skins it brings an inflammation worse than freckles.'

He had final words of caution. The Hoop Tree or Bead Tree (*Melia azedaracta*) was imported from the Americas, but with a view to having hoops grown in this country. But it was either too cheap or not found to answer the intention, as it is seldom if ever made use of by the planters.

> Some people here think this plant is poisonous but I cannot think so as horses eat the berries without injury and even fatten on them. This by the by is a good mark to judge of plants or fruits and I

have made it a rule never to taste any leaf or fruit that are avoided by cattle or insects.

Even pigs were not immune from observation in their eating habits. He noticed in the West Indies that when the hogs rooted up tubers of the bitter cassada plant, covered in mould and mud, there were no ill effects, but when fed with the same plant washed they became ill and died. If this bitter cassada was mistaken for the common cassada, which was eaten in copious quantities by the natives, the immediate response was to order the patient to drink large quantities of warm muddy water. The cure appeared to work.

Between times Wright commented briskly on the outbreak of fever amongst troops at Walcheren: 'It is none other than the endemic fever of marshy countries and . . . the winter will put a stop to it.' Although he claimed to know almost nothing of the local agriculture of Scotland, he was elected a member of the Agricultural Society of Strathearn, and asked by Sir John Sinclair, responsible for the Statistical Accounts of Scotland in 1791, to write a paper on the potato. He concluded that potato bread would cost a third less than bread made entirely from corn.

Meanwhile in Crieff his nieces were embroiled in helping a needy female society and raising money through subscriptions and balls. His niece reported scenes of great wretchedness, a 'Mrs Weems in particular lying in a corner of a dark garret on straw, some dirty blankets, but neither sheet nor cape, and such a state of filth as the house was cannot be described. Sandy McCulloch's family in a sad state. Him blind, the daughter daft, the wife and son ill of fever.' As ever, Wright held an opinion, and proffered that he was in agreement about the wretchedness of the lower orders:

From my many conversations I have had with sensible Guinea Negroes, I think they change their climate and condition for the better. They described their country to be hot, sultry, and in many places unhealthy; their habitations as temporary and miserable, infested by noxious animals, and surrounded by hostile nations, so that their lives and properties are perpetually in danger. They are brought to a fine healthy island, where, in a little time, they find themselves quite at home, in safety and under protection. The Negro is supplied with every necessity of life both in food and clothing. He has a good house and proper utensils. When at length he is put to work, it is proportioned to his strength. The heat of the sun is so far from being hurtful, he takes delight in it. This too, is precisely the case with his descendents.

His biographer, a local Crieff worthy called Mr Turnbull, who had been appointed by Wright's nieces to write up his memoirs, and who offered negligible asides about Wright himself, was moved to offer a rare opinion. Turnbull was openly puzzled by why men who had spent much time in areas populated by slaves remained convinced in the system. Why did they arrive in these islands horrified by slavery, yet leave and continue for the rest of their lives satisfied with the practice? Perhaps Wright had convinced himself that the rumours of Africa might be true, that native Africans sharpened their teeth purely to enjoy 'the inhuman banquet of bodies of their captive foes'.

Regardless of such speculations, Wright's contribution to botany and plant collecting was extraordinary for a man who seemed to desire nothing more than to have his feet firmly at the fireside of a modest, comfortable Perthshire house. All he appeared to have wanted was to be a country doctor. He sought no adventures in life, but somehow landed in the midst of wars and settled down for many years on tropical islands. The remains of his massive collection of West Indian plants still rest at the Royal Botanic Garden in Edinburgh. His many theories on the cures for such illnesses as diabetes were the subject of discussion in London and North America, and his correspondence with the eminent men of the day, such as Sir Joseph Banks, reveals a well-travelled man who loved to bask in the friendship of the great men of the day. Would he have been thoroughly bored if life had served him a rural practice in Perthshire? Probably. The Royal Navy offered him uncomfortable situations in dangerous areas of the world, but without the navy, the plants of the West Indies would never have been catalogued so efficiently and thoroughly in the eighteenth century and ended up in Scotland.

When Wright died in 1810, John Fraser, fifteen years his junior, was languishing ill in London and was to die a few months later. While Wright was able to spend his last few years in a much enjoyed retirement, with the luxury of sauntering around the Trossachs, and of issuing advice to would-be travellers in the tropics from the safety of his native Crieff, Fraser's final year was quite the opposite.

CHAPTER 5

John Fraser (1750–1811)

CATHERINE THE GREAT AND THE CAROLINAS

———

John Fraser's elegant portrait depicts him as a foppish dandy, with his coiffed hair, white stock and velvet-collared frock coat. There is a touch of the dilettante about him, of one who has chosen the gentlemanly hobby of gardening, and is languidly fingering a flowering stem with one hand, while the other props up a polished gardening tool. Fraser has a handsome, fine-featured face, high cheekbones, and a steady gaze. It was an altogether convincing pose, but a complete ruse. The Fraser clan emblem is a strawberry growing out of a dung heap; was it a telling indication of the future of this lad from a croft?

Certainly there were Frasers aplenty in the Inverness area of his birth. One well-known scion had already spawned sons of varied talents in the previous generations.

But our John Fraser, born in 1750, came from lowlier stock. He was probably the fifth child of Donald Fraser, born in the area to the south of Inverness into a crofting family. He was christened in the parish of Kiltarlity on 14 October 1750 and the entry in the parish register reads 'Donald Fraser of Tomnacross, had a child baptised named John, witnessed by the congregation'. John Fraser would have absorbed from the cradle a culture of self-sufficiency, allied with an ability to turn his hand to many skills needed for survival. Such a background in small-scale tenanted farming was a hardy training for life and nurtured initiative. The land itself could not support a family and family members would turn to sideline occupations to make ends meet.

Crofting sons knew that they would be at least thirty before they could inherit, if their father lived to the average age of fifty-five or so, and this applied only to the eldest or most able son. So John Fraser, the fifth child, would always have known that he would have to make his own way by different means. Life on a croft required a multitude of hands-on skills: farm husbandry, building the family house or farm steadings, thatching, stonemasonry, hunting and trapping, fishing, reading the weather, watching the crops for diseases, candle-making, and preserving of foods would have been learned for survival. Dry-stane dyking might bring in some extra money, or, with Inverness supporting a buoyant trade in flax with the Baltic areas, weaving with flax or wool might take place within the house. His father also appeared to be forward-thinking in agriculture and they lived in an area in which traditions of plants and their uses were well known to various Fraser relatives, some of whom were so talented and skilled in the realm of botany for medicinal use that they were granted free land in return for growing relevant medicinal plants. An early appreciation of plants and their uses might well have been absorbed by young John Fraser.

He would also have been bilingual: Gaelic was spoken at home, broad Scots further afield, and with a couple of languages on the tip of his tongue, his ear would have been easily attuned to picking up more. Added to this was the influence of the Free Kirk and Calvinism, which preached that every man was equal in the eyes of God. The class system in Scotland was less rigid than in England, and education was honoured. The house would have had its Bible to hand, and story-telling, singing and music were part of everyday life. All in all it was an upbringing which couldn't fail but to instil self-confidence.

Fraser would have set off as a 'journeyman' following an apprenticeship to a weaver. As he journeyed down through the land, picking up work here and there, steadily veering southwards, he would keep a sharp ear open for ships bound for London. By the time Fraser had reached his early twenties he was not only established in London, but had embraced the opportunities of a new consumer economy, of a nation of shopkeepers. He appears to have been a hosiery and linen draper. He was an opportunist and an optimist, and things appeared rosy. Life, however, was not altogether set fair.

Consumption (tuberculosis) seems to have stalked Fraser, no doubt made worse by living conditions within the capital. Most likely he lived above his shop, and the restrictive life behind the counter, indoors and in cramped, dusty conditions, when he had been brought up with the open spaces and fresh air of the Highlands, might have been devastating to his health. Despite this setback and his lowly background, he had somehow befriended or impressed Admiral John Campbell, whom Fraser's son later described as a friend of the family. More likely he might have been a customer in his shop. Tellingly, the shop was situated in Paradise Row, close to the Chelsea Physic Garden. Campbell had circumnavigated the world with Captain Anson on the *Centurion*.

On Midsummer's Day in 1778 Fraser was married to Frances Shaw, and at least two sons were born: John Thomas in 1780 and James in 1782. Just after the latter's birth, Fraser set off for Newfoundland – where John Campbell was a Vice-Admiral and had just been appointed as Governor and Commander-in-Chief – to recuperate from a bout of tuberculosis. How Fraser managed to manoeuvre himself from standing behind a counter in London to Newfoundland is a total mystery. Perhaps he was taken there as a clerk or in some capacity as a linen draper. What is released from Fraser's later accounts is that he spent the three years of Campbell's appointment roaming the countryside, and developing an eye for botany.

His was a sudden metamorphosis into a plant collector. The next we hear of him is his dispatching his finds to William Forsyth at Chelsea Physic Garden. He commenced a correspondence with Sir James Edward Smith of the Linnean Society that came to include William Aiton, King George III's gardener at Kew. All this was made possible by easy access to regular ships sailing from Newfoundland to England. Ironically, he might also have come across his native heather during his plant hunting. Stories abound of heather accidentally growing in the area due to seeds being dislodged from the boots of the Black Watch regiment when they moved billets from New York to Halifax, Nova Scotia, in 1783, the year following Fraser's arrival.

Whether he returned to England in the interim is hazy, but by 1785, he had set off for North and South Carolina, the hosiery trade forgotten. An unsubstantiated report in Loudon's *Arboretum et Fruticetum Britannicum* claims that he went to Charleston, South Carolina a year earlier. He sent seeds and plants from the Carolinas on to a London nurseryman, who claimed they were 'dead on arrival, or common and not very saleable'. Returning to England, Fraser launched an unsuccessful claim against the nurseryman in the courts. Undeterred, and presumably with his ever-supportive family behind him, he returned to the Carolinas. This time he arrived in Charleston on 20 September 1786.

Within another couple of months he had introduced himself to André Michaux, a Frenchman with a penchant for adventurous botany, a generous grant from the French government for the establishment of nurseries in New York, and a shopping list of plants to be sent over to France. Michaux arrived in New York in October 1785, immediately establishing and organising a nursery there in order to grow plants which were on order for the park at Rambouillet, near Paris. There, they were duly received and distributed by the Abbé Nolin. Fraser soon assumed he was Michaux's honoured close friend.

The son of a well-to-do farmer, Michaux was four years older than Fraser, and had studied botany and the craft of gardening at Versailles. Exploring the East with the sponsorship of a member of the French royal family, he had journeyed as far as Afghanistan. There, he was attacked and left for dead before being rescued. While convalescing he filled his time by compiling a French–Persian dictionary. By many accounts, although possibly all based on his own version of events, he was well able to take care of himself and showed remarkable courage and resilience. He had also picked up camellias, mimosa, and the sweet olive on his way.

Fraser not only thought he had found the ideal travelling companion, but that he could surpass Michaux's abilities to collect plants. Fraser immodestly declared that he 'at once took up a determination, which may be thought rather presumptuous, of endeavouring to excel Mr Michaux, or at least share with him the honour of extending the knowledge of my countryman over the vegetable kingdom in this part of the world'.

On 11 November 1786, not the best time of the year, seasonally, for plant collecting, Fraser bought a horse and chaise and was joined by a Dr Porter and a 'young Mr Brooks', who made his way to Savannah in Georgia by sea, while Porter and Fraser rode. Porter was meant to share the expenses of this trip. Fraser employed Brooks for his expertise in collecting and preserving birds and insects. Leaving Porter at Savannah, Fraser set off by himself for Sunburry (now Sunbury) about forty miles south. In the aftermath of the War of Independence from British rule in 1775, Georgia, named after King George, was a breakaway state from British rule. Fraser found lodging with a Colonel Elliot, there being no inns on the road. Despite his nationality, he was well received, and his host's son entered into the spirit of plant collecting. They set off in tandem the following morning. Ahead was a glorious scarlet azalea, described by Fraser as a 'Scarlet Ezilia', spelling not being his strong point. As this was autumn, it would have been the brilliance of the glowing, vermillion autumn leaves which caught his eye.

Arriving at a pub at Sunbury, one of the important coastal towns of the era, he booked in at a local inn, accepting an invitation from the landlord to go on a hunting trip with seventeen other – presumably new – companions. Leaving his precious plants in the care of his 'old friend' Porter who, with Brooks, was to be in charge of tending his collection, and remain behind in Savannah, Frazer and the seventeen others packed themselves into a log canoe, carved from a single tree trunk, and he set off to hunt plants while his new friends hunted game.

The first night was spent in heavy rain, under the partial shelter of a cedar tree. Breakfast was oysters and hominy. Exploring the island of Saplo alone the following day, Fraser became lost and finally climbed a tree to get a bird's-eye view of where he might be. Eventually, soaking wet and exhausted, he was found by a search party, only in order to spend another night in pouring rain without shelter. To revive him, gin and pepper were poured down his throat, although he reckoned that if he had been force-fed with arsenic it would have been merciful, feeling as wretched as he did. But it was a miraculous cure, and despite not having taken his coat off for ten days, he was up and about the following day and even accepted another invitation to explore the further islands of Frederico and Blackbeard. Before heading off again, he was surprised to encounter Porter, whom he had thought was dutifully partnering Brooks in Savannah. Porter assured him that Brooks was remaining at Savannah, caring for the plants alone, and awaiting Fraser's return. Perhaps alarm bells should have rung. Why had Porter simply left in defiance of Fraser's instructions? But Fraser was oblivious. Off he set on the second expedition, eventually arriving at the Turtle River and dispatching his collection on a boat for Savannah. This time McIver, a local resident who had acted as guide, fell ill, and Fraser left him to nurse himself in a wooden hut in the forest, while he made for Sunburry, just south of Savannah.

To his horror he found Brooks gone, with all his plants and without leaving his share of the expenses. Fraser set off in pursuit, arriving just in time to find Brooks already ensconced on a London-bound ship, and challenged him about his conduct. Brooks excused himself by retorting that he would not stay with such a 'disagreeable' young man as Porter and had taken himself off, along with Fraser's entire collection of plants and seeds, to travel home. Whatever the reason for the serious disagreement between Brooks and Porter, and their clear disregard for Fraser's instructions, Fraser wrote not another word about this episode. He must have been able to sort out the situation, and retrieve the seeds. He never mentions either Porter or Brooks again.

Ever the gentleman, Fraser confined himself to the bare account in his letter to William Forsyth at Kew. He offered no criticism but, with a masterly stroke of irony, included a large hornet nest for Forsyth. For Sir Joseph Banks there were a few gentians, and a new laurel, which, Fraser added, was numbered as No. 18, and which he thought was a new variety. Fraser added for good measure that he thought it would be 'very unfortunate' if he had not added a few which were new, and on the strength of supplying such wonderful new plants, he hoped that a modest financial advance might be forthcoming to aid his search. As a postscript he wrote that the 'paper medium of this state [i.e. the local currency printed by the State of Carolina] goes nowhere else', in other words that the Georgia currency was only exchangeable in Georgia. Fraser was ever to feel at a disadvantage to André Michaux, whom he correctly perceived was better financed and whom Fraser could not resist picturing at every opportunity as a man of lesser fortitude.

Fraser was at pains to drop into his letter that he and Michaux were due to explore 'Indian country' together. The description he gave quivered with danger, excitement and the unknown. Hanging in the air was a possibility they would not return, but it emphatically would not be himself, John Fraser, who would let the side down. With tongue firmly in cheek, he relayed an incident in which Michaux could not resist coming on board the *John* to view the plants Fraser had found, which were about to leave, presumably with the errant Brooks. A storm blew up and the Frenchman was almost too terrified to leave the relative safety of the large ship to go ashore that evening in the small rowing boat. The alternative would have been to stay on board bound for England. Fraser managed to persuade him.

By now it was January 1787, and in the April they set off together for the Indian territory. Michaux's horses were stolen between Savannah and Augusta and as the hapless Frenchman searched to retrieve them or procure replacements, Fraser set off in advance for country he was to describe as the 'Italy of America', a title which can only have been based purely on the whimsical drawings brought back by those on the 'Grand Tour', which he might well have seen in London. Contrary to Fraser's hints that venturing into 'Indian country' might be a life or death experience, the very few native Cherokees he encountered were peaceful and friendly. Carrying satisfyingly full botanical tins he returned to Charleston. Here the ever-social Fraser met and formed a close friendship with a local explorer, Thomas Walter, who was in the final stages of his magnum opus on the plants of the Carolinas. At Fraser's instigation, Walter enlarged this already bulging collection of 640 species with Fraser's 420, and promptly died in January 1788. Fraser packed up the manuscript – he appears to have agreed to undertake to publish the work in England when he returned – along with 30,000 plant specimens and many boxes of living plants and sailed for home, arriving in March 1788.

He was now virtually penniless. His trip away had lasted nineteen months – a substantial time with no income. He gave away few of his plants, some of which should have been dispatched to his subscribers, who would only have contributed to sponsoring his voyage in the expectation of being allocated substantial numbers of plants in exchange. In direct contradiction of what is likely to have been the agreement, Fraser lost little time in selling the plants himself, offering them first to his subscribers, many of whom were understandably irritated and felt cheated. One, Dr Lettsom, wrote to a friend Humphry Marshall on 2 February 1789 that 'Fraser, to whom a few of us in London subscribed an annual sum, has not answered our expectations'. Fraser did not name the disgruntled clients himself, perhaps presuming that charm and fast talking would win through. However, as Joseph Banks was one of the hopeful clients, Fraser had made a major blunder. He was not in a position to snub or cheat such a wealthy and famous benefactor.

However, irrepressible as ever, he plunged into publishing the late Thomas Walter's *Flora Caroliniana,* somehow bearing the cost himself. Inscribed on the frontispiece was this dedication: 'To Thomas Walter, Esq., this plate of the new Auriculated Magnolia is presented, as a testimony of gratitude and esteem, by his much-obliged humble servant, John Fraser'. While Fraser did not appear to claim authorship, he certainly ensured that his name appeared in a prime location within the book.

Money may not have concerned Fraser too much, though, as he felt he had a financial trump card up his threadbare sleeve. He and the hapless and now deceased Walter had been certain that they had stumbled across a seed that might prove the gold mine which would set them on a secure financial footing. This was *Agrostis perannaus* or *cornucopiae,* a lacklustre grass of lax habit and forgettable attraction, the smallish seeds of which were eaten by the native Cherokee, either dry or mixed with water to form a mush. Although this did not sound a great botanical find for the great gardens of Europe, he was convinced that as an easily grown cereal crop, the grass would provide a new, cheap food product. However, he encountered little enthusiasm in England. Instead of nurturing and consolidating his botanical finds in his London nursery, and repairing and cementing relationships with his subscribers, all of which actions would have kept his finances and business buoyant, he set off for France to see if the French would buy it, thereby making his fortune. His giddy optimism was to be short-lived.

Arriving in Paris at some point in late 1788 or early 1789, he is reputed to have met the future American president, Thomas Jefferson, who was an envoy there. Jefferson was touring Holland in March and April 1789, and left France on 22 October of that year, just after the outbreak of the French Revolution on what became known as Bastille Day, 14 July. Perhaps Fraser viewed Jefferson as a splendid contact. Jefferson had a farming background, and was in the midst of building Monticello, his mansion in Virginia, creating both a house and garden in which he had a lifelong and

100 SEEDS OF BLOOD AND BEAUTY

passionate interest, and might well have been interested in a plant explorer, especially one freshly returned from near his home state, Virginia. But if Fraser elicited any support or further contact with Jefferson, it is never mentioned, and with Fraser's ill timing on the eve of the French Revolution, his get-rich-quick ideas vanished. Yet again he returned home empty-handed.

He made at least two trips to the same areas of America between 1789 and 1795, although a visit to Jefferson and Monticello never seems to have materialised. How he financed these trips is vague in the extreme. Only a few facts are certain: gone was the reputedly frail young man who had required four years of recuperation in Newfoundland; and gone was the London hosiery shopkeeper, replaced by a nurseryman with a business in London, a 12-acre nursery of his American plants close to present-day Sloane Square, in London's Chelsea. 'American' gardens were a new and exciting vogue, and appealed greatly to men of means, who took to natural landscaping and cleared areas of their vast estates to plant up conifers. Fraser felt at the forefront of this new fashion, and accordingly zigzagged back and forth to North America roughly three more times. But there was also an additional incentive for his foraging.

In 1796 he suddenly pops up in Russia, taking a supply of plants and successfully selling them to the legendary queen, Catherine the Great. Catherine the Great was an international traveller with a generous purse. Much has been written about her prodigious appetite for lovers, but her appetite for artefacts and ideas from around the known world was as all-encompassing. She commissioned the first landscaped park in Russia in 1770, employing John Bush and Charles Cameron to design bridges and buildings. She ordered an album of the layout of Hampton Court Palace gardens to be specially drawn for her in 1780 by John Spyers, surveyor to Capability Brown. She snapped up a collection of botanical drawings from the estate of Dr John Fothergill, physician, philanthropist and botanist, who had died in 1780. Catherine bought them through Sotheby's in the year following his death.

Dr Fothergill was a successful doctor, but he also found time to cherish his botanical garden at Upton House, Essex, now West Ham Park, East London. This became Britain's first alpine and wilderness garden, the greenhouse and hothouses filled with 3,400 exotic plants, tended by fifteen gardeners. He ordered artists such as Georg D. Ehret, John Abbot and John Miller to depict 611 of the plants in his gardens. It is this collection which Catherine bought and which would have given her a taste for unusual and, to her eyes, new plants.

Having viewed the paintings of these plants, Catherine would have wanted the living specimens, and there would most probably have been a flurry of activity to supply her wishes.

How Fraser came to be the man for the job is unclear, but he was certainly moving in the right circles. There were many connections. Dr Fothergill was a patron of the American William Bartram, who was travelling and indeed finding similar plants to those gathered later by both André Michaux and Fraser. Bartram's specimen plants were purchased by Sir Joseph Banks from the estate of Fothergill. Fraser, however, was responsible for publishing *Flora Caroliniana* in 1788, while Bartram was a little behind as his published travels appeared in 1791. Fraser was available, and contactable with greater ease, being at the time on the European side of the Atlantic. Why wait to contact the American, Bartram? Catherine appeared anxious for new plants, and Fraser might just have been residing in London when the queen came over to England. Both the Russian Ambassador to London, Semen Romanovich Vorontsov, and Catherine's Scottish doctor, Dr John Rogerson, may have aided Fraser in sealing the deal.

However it came about, Catherine commissioned Fraser to go to America and collect plants just for her. In 1799, armed with a commission and accompanied by his son John, he set off for the familiar stamping grounds of the Carolinas. He was understandably revitalised, buoyed up with the honour of such an assignment. Success, both financial and public, seemed within his grasp. Fraser's eternal optimism looked as though it was about to pay fat dividends. Although previous experience appeared to have taught him little about business, this time he set off from Britain with one of his Russian finds packed in his bags for America. He took with him the black and white Tartarian cherries which he had already introduced into Britain. These came to be known in the United States as Fraser's Tartarian cherries, and are still known by that name and prized over many parts of the continent.

Confidence boosted, father and son appear to have set out for Tennessee, Kentucky and Ohio. The Frasers left almost no account of this journey, save of a defining moment on the Bald Mountain, or Great Roa, which John Fraser eulogised in an account published in the 1852 *Cottage Gardener*:

> I shall never forget so long as I live the day we discovered that plant (*Rhododendron Catawbiense*). We had been for a long time travelling among the mountains, and one morning we were ascending to the summit of the Great Roa, in the midst of a fog so dense that we could not see further than a yard before us. As we reached the top, the fog began to clear away and the sun to shine out brightly. The first object that attracted our eye, growing among the long grass was a large quantity of *Rhododendron Catawbiense* in full bloom. There was no other plant there but itself and the grass, and the scene was beautiful. The size of the plants varied from seedlings to about two feet in height, the habit being evidently diminutive, from the high altitude at which they grew. We supplied ourselves with living plants, which were transmitted to England, all of which grew and were sold for five guineas each.

Tired, damp, almost certainly hungry, probably uncertain of their exact whereabouts, the sight must have been like a fat London bank account fluttering in the breeze, with the happy addition of Russian roubles as an addition. Rhododendrons were unknown in Britain as living species at that time, except through paintings and sketches. It was not until almost fifty years later that more rhododendrons came to the West, brought from the foothills of the Himalayas by Joseph Hooker, the son of Sir William Hooker of Kew. Fraser's discovery bears pale pink flowers, which today might look insipid compared with the riot of colour that originates from Asia, but the delicious, delicate, frilly flowers caused a sensation back in London, and possessed that other vital characteristic: despite their delicate appearance, they are reliably hardy in most of Britain.

Rhododendron catawbiense must have imbued the pair with zest for more adventure, although Fraser senior appears to have been the more devil-may-care of the two. He might have been forgiven for feeling he deserved to see more of the world and allowing himself an adventure or two before the trip home. He had reached his fiftieth birthday, despite being visited by the dreaded tuberculosis twenty-five years earlier when death would have seemed just round the corner.

Instead, here he was, on top collecting form, with a son to be proud of alongside, a box of plants in transit to London and a secure financial success awaiting his homecoming. It was a moment to savour. Prudent men would have packed their bags with the trophies firmly within their grasp and headed for home. Instead the Frasers decided to turn south and head for Cuba. There was a minor problem in their way, as Britain was at war with Spain, and Cuba was in Spanish hands. This was solved by the simple expedient of acquiring false American passports.

Armed with the appropriate paperwork, they set off and, perhaps unfairly, but almost predictably they immediately ran into trouble. Storms wrecked their boat on a coral reef eighty miles from Havana, possibly those associated with the Cayos islands. For six days the Frasers, along with sixteen of the crew, clung to the remains of the boat, eventually being plucked out of the water by a passing Spanish vessel. If they had not anticipated being shipwrecked, they certainly had not bargained on losing their passports so soon and being rescued by a Spanish ship and taken to a Spanish island with no means of verifying their 'American' credentials was, to put it mildly, inauspicious. They landed at the port of Matanzas. This time their luck held, and the American consul swallowed their tale, allowing the nearly destitute pair to proceed overland to Havana.

Good luck struck a second time. Botanists and travellers Humboldt and Bonpland were also exploring the area, and as fellow botanists the Frasers were received kindly. Fraser, ever one to enlarge his circle of acquaintances, could well have become known to the Frenchmen during his years exploring. Humboldt offered the Frasers financial aid and free lodging in his house, but felt justifiably uneasy. Much as he wished the Frasers no harm, he was in a tricky position: should the Spanish find out he knew they were British and was sheltering them, he could at the very least be deported without having rooted out the plants he desired.

In confidence he told the governor of Cuba, Marqués de Someruelos, that he knew the Frasers were British. It was a wise move on Humboldt's part. The governor was minded to turn a blind eye, declaring that 'Though my country is at war with England, she is not at war with the labours of these men.' He supplied them with passports to ensure their safe passage round the island.

Fraser was relieved, and happy to find that his son so impressed Humboldt that he invited Fraser junior to accompany him to Mexico. Young Fraser, however, was wary of travelling in another Spanish country, spoke no Spanish and in any case was ready for home. Instead he packed up their own plants, was entrusted also with Humboldt's boxes and set off for Charleston and home, setting foot on British soil in June 1801. John Fraser senior remained in the Carolinas for another season, and set off for England in 1801. This time his ship only sprang a serious leak rather than sinking, and he had to spend some time in Port Masson (now Nassau) in the Bahamas before arriving back in Chelsea. Here, he was met by a body blow.

Although Catherine had died in November 1796, Fraser had understood that he was still very much the collector of the North American plants desired by the royal house of Russia. Catherine's successor, Tsar Paul, and his wife, Empress Maria Feodorovna, wanted his plants, or did they? Tsar Paul was assassinated in March 1801, and his successor, Alexander, refused to honour the agreement with Fraser. Undaunted, Fraser set off for Russia, eventually travelling back and forth a few times. He later put forward the claim that he had been commissioned to bring back plants for the Russian Court, and this was disputed. Fraser tried every avenue, and pressed every contact he could think of, to further his claim. At some stage he took with him letters from Lord Whitworth, the erstwhile Ambassador to the Russian Court and then President of the Linnean Society, along with another separate request from the Linnean Society, asking the Russian Court to give credence to his cause. While Fraser was in Moscow, the might of the British diplomatic mission in the form of the British Ambassador, Sir Borlase Warren, also tried to intervene. Fraser spent an entire summer and winter there progressing from office to office, courtier to courtier, hanging around waiting for days and weeks for an interview. He must have felt all his hopes seeping away. Bitterly disappointed, ill, and confined to bed for several months, he was additionally at the mercy of the approaching winter and biting cold.

Finally, he decided to return home. Only the widowed Empress Maria, wife of the late tsar, had taken up his cause, but the best that seems to have come of this support was her gift of a diamond ring, and some very small compensation that appears to have been paid. In the meantime, Fraser had one more bright idea up his sleeve. His sister, Christy, who lived with the family in London, took up his venture with straw hats, using the leaves of *Coccothrinax miraguama*, which he had discovered in Cuba. There Fraser had watched hats being woven without the need for stitching, and was convinced this was a splendid new, money-saving method of constructing hats. But despite Christy's efforts, even with passing interest from Queen Charlotte, who apparently did purchase a few, this project too ended in financial failure.

By now it was 1807. Fraser had wasted several valuable years under the assumption that his Russian fortune was around the corner and that his prowess as a collector was to be honoured. Apart from his nursery, apparently run by one of his sons, and a long-suffering family, nothing remained of his dreams.

Fraser was a generous, if too trusting, man. He was impulsive, clever, knowledgeable, imaginative and possessed of great charm. His life is a story of successful international networking. By his fifties his circles of acquaintances encompassed Europe and eastern America. Not bad for a lad from Inverness. There was just one blip, extraordinary in a man who appeared to have acute antennae for making and retaining friends. Fraser had managed to upset that bastion of the plant collecting establishment, Sir Joseph Banks. This seems to mean that he did not see himself as Banks' courier, but rather a plant collector of note to be respected in his own right. Banks, not surprisingly, had not appreciated this upstart's view, used as he was to young collectors only too willingly donating selections of their finds to his ever growing collections. Put simply, Fraser had tried to sell his plants to stave off bankruptcy instead of offering them to Banks, probably as previously agreed.

Fraser was fifty-seven, and he had only one possibility left. Faced yet again with desperate financial problems, he decided to take the only course he knew to rescue the situation, and that was to return to plant collecting in America. Accompanied by his now thirty-year-old son John, he sailed in 1807. Fraser senior, as ever when out of sight of the humdrum life of a businessman, found a second wind and they found many more plants for his nurseries in London. This time father and son stayed longer, with John returning with the booty in 1809. Yet again, his father decided to prolong his stay for another season, even visiting Cuba again, before returning to Charleston for the journey home. Just a few weeks before sailing, however, he fell heavily from his horse, breaking a few ribs. But he was far from medical aid and with his injuries still fresh and unhealed he sailed for home. On arriving in Britain in the spring of 1810, he was an invalid, and bedridden.

He lingered on for almost a year. When he realised that he was dying, he made one request, that Sir Joseph Banks, from whom he had been estranged all his life, would come to his bedside. Banks acquiesced. Their conversation must have ranged over many issues, but Sir Joseph left him as a friend, and Fraser died shortly afterwards, in April 1811.

Fraser was the man who, literally, brought America to Britain. In a shrewd marketing move his Chelsea nursery was named 'The American Nursery', and had no known rivals. The trees and shrubs he packed up in the soil of South Carolina and Georgia created a zing in Britain. The rhododendron was the first of many plants to arrive, as well as 450 species of American oaks, often with leaves that glowed scarlet and orange in the autumn, which must have been a marvel to behold for Britons more used to mellow golds and yellows. For good measure, he also managed to bring back an oak from Russia, *Quercus castaneifolia*, with long, elegant graceful leaves.

Flowering, flamboyant trees arrived too. The magnolias he brought back produced profuse, large flowers, the size of a gentleman's neckerchief. Flowers which seem so common now in suburban gardens arrived, including evening primrose, daisy-flowered species such as *Gaillardia* or blanket flowers, *Erigeron aureus* (fleabane), *Collinsonia*, a popular annual with purple-pink flowers, and *Coreopsis*, known popularly as tickseed or tickweed. Shrubby hydrangeas, *Andromeda* (bog rosemary), nodding flowers like *Allium cernuum* and pretty ground-covering varieties of *Phlox*, such as *P. stolonifera* which creeps over the earth and produces delicate white flowers, arrived too. The curious purple flowing *Liatris* which flowers from the tip of its flowering spike downwards, in contrast to most flowers, and the *Viola pedata*, the simplest of the pale mauve violas, all came from Fraser. He did succeed in having a species named in his honour, *Frasera speciosa*.

His son John seems to have learned from the financial failings of his father. He went exploring in 1817 to the same areas as his father's expeditions, but journeyed further south into eastern Florida. Returning, he successfully auctioned off his plants, and with the proceeds established a nursery called 'The Hermitage' by Ramsgate, retiring in 1835 and handing over to William Curtis, whose father was the proprietor of the *Botanical Magazine*.

CHAPTER 6

Archibald Menzies (1754–1842)

ALL AT SEA WITH VANCOUVER

Archibald Menzies (pronounced 'ming-us') was born either at Weem, close to where his father was the gardener at Castle Menzies by Aberfeldy in Perthshire, or – more likely, as this was where he spent his childhood – in the parish of Newhall, or Sticks (now confusingly spelt Stix), which is across the River Tay and around five kilometres from the Z-shaped bulk of the castle. Although the splendid Wade bridge – perhaps the finest example of its type in existence – was built over the Tay at Aberfeldy in 1733, making the crossing of this fast-flowing river possible, it is likely that Menzies' father would have crossed over by ferry immediately below his house. The ruins of the family's farm buildings still sit there up on the hill.

The laird across the Tay, on the north side, was Sir Robert Menzies, the third baronet. His family were keen planters and great enthusiasts for trees. He planted a larch from Tyrol, the first of its type to be planted in Scotland. He was the owner of large estates.

When Archibald Menzies walked to school he saw changes happening all around him, and it was all due to the introduction of sheep farming amongst tenants around the time of his birth. It would be another thirty years before landowners, seeing the value to themselves of sheep, would begin wholesale eviction of tenants from their lands. Meanwhile, the sheep brought some prosperity, although there was still much to learn, notably that to achieve good quality fleeces would require much care. The great alluvial plain of the River Tay supported rich farming. The dragnets in the fifteen-mile long Loch Tay yielded up quantities of salmon. Salmon was almost free, the cheap food of the age, and servants were often quoted as refusing to consume it more than five times a week. Close by, the lochs in Glen Lyon and Glen Roro were reputed, in the *Statistical Account of Scotland 1791* for the Parish of Weem, to allow at least one keen fisherman to 'catch 200 trouts in a day, weighing from 4 ounces to a pound weight a-piece'.

Although the rebellion of 1745 led by Charles Edward Stuart had left a legacy of disillusionment, it was a time of much advancement around Aberfeldy. Supplies for the local area, previously brought in on horseback, now flooded in. Supplies of salt, iron and tar also arrived more easily. Other improvements were taking place. Lime was under production locally, and rapidly effected improvements to the quality of the soil, although fuel was in short supply for the high heat required for limekilns. Turnips had appeared, being laid in drills and 'most carefully hoed'. Prior to almost any recreational use being made of the mountains for climbing, their beauty is mentioned in the *Statistical Account*, as well as the great variety of local alpine plants that are rarely seen in other parts of the country.

Overall, to be born in this area at this time was fortunate for Archibald Menzies, in comparison with John Fraser, who had first seen the light of day just four years earlier near Inverness. Perthshire was beginning to enjoy a period of stability and advancement, raising the standard of living. Further north the land was more rugged and wild. Widespread inoculation meant that smallpox now carried off one in two hundred instead of one in seven. Weem and the surrounding area were relatively prosperous and there were few poor in the parish. Menzies would have spoken Gaelic as a first language, although he would have been easily conversant, almost bilingual, in English, a great advantage for learning other languages. As a gardener, his father was not dependent on an agricultural tenancy for his livelihood, although he would have cultivated the land around the family croft. The coming of the sheep brought differences to crofting life. Previously folk had migrated high up into the hills with their cattle for most of the summer months, returning for the harvests. Sheep required less movement from home, so there was little reason to keep children away from school for lengthy periods and schooling was therefore more continuous.

Perhaps the most positive and lucky factor of all for Archibald Menzies was that the garden of Sir Robert Menzies was in the process of becoming adorned with ornamental trees of many types. By the early eighteenth century the trees surrounding Castle Menzies ranged from plane trees, to Spanish chestnuts, silver firs, oaks, spruces, ashes and beeches. They were so robust and generously sized that they were carefully measured and noted. The walled kitchen garden, settled into a steep south-facing slope, was terraced and grew copious quantities of fruit and vegetables. Both Archibald and his elder brother William served time in the garden before both walking to the Royal Botanic Garden in Edinburgh, Archibald by now being around the age of twenty. The Menzies brothers, including a third, James, were unusual: all felt the need to flex their wings. Few others of their schoolmates left the area.

From around 1774 to 1778 Archibald was a well-known face in Edinburgh's Botanic Garden. His brother William was already established there as a gardener. Both came under the influential and kindly wing of Dr John Hope, the professor of botany at the University of Edinburgh. Archibald's first foray into plant collecting came in 1778 when he was twenty-four, in a vicinity close to home. Presumably with the encouragement of Hope – he could never have afforded to take time off without his employer's sanction – he toured the hills of the Scottish Highlands and the Hebridean Isles and without doubt his local knowledge must have stood him in good stead. All those miles walked around Perthshire had fine-tuned his appreciation of local flora. He would have been more than able to pick out an unusual find in what others might have glossed over as acres of heather or grass. His finds of rare plants were added to Lightfoot's *Flora Scotica*. Armed with this success, he studied botany under Hope. How he funded this is unclear: perhaps he could afford to pay a small subscription; or sometimes townspeople could sit in on lectures free at their local universities; or perhaps Hope saw a promising young man, and waived a fee. It was Hope who also encouraged him to study medicine.

He was entering a world far distant from the supportive and egalitarian atmosphere of the Edinburgh Botanic Garden. It was only twenty years since William Wright left Edinburgh to take his chances with the Royal Navy as a surgeon, and Menzies came from much the same artisan background. Both had fathers who aspired to greater things for their sons, but neither had the cash to allow them to live anything like the gentlemen who surrounded them. Walter Jones, an American student, wrote to his brother in 1776:

> The students here, dominated by medical, may be referred to three ranks or orders,
>
> First, the Fine Gentlemen or those who give no application to study, but spend the Revenues of Gentlemen of Independent Fortunes.
>
> Secondly, the Gentlemen, or Students of Medicine strictly speaking, these live genteely and at the same time apply themselves to study.
>
> Thirdly. The Vulgar, or those who, if they are not indolent, are entirely devoid of everything polite and agreeable. I believe you will not doubt for a moment with which of these orders I ought to associate.

It was as a gentleman that American student Walter Jones in 1776 penned the above to his brother back in Virginia. Curiously, as the American War of Independence raged, there was a surprising number of students from the very recent 'colonies across the Atlantic'.

Jones' three-year stay 'as a gentleman' managed to bankrupt his brother back home, as he ruefully justified, through the expense of living in Edinburgh. Having intended to manage on £90 per year, instead he ended up spending £118. It was not, of course, all work. There was leisure time for those who could afford it, and many had allowances of £150 or even in an extreme case £200 per annum, and if they were not playing cards, or gambling, they attended plays. In fact, the Royal Medical Society was so incensed at students who regularly missed lectures, that they invoked the right to level an accusation that they had been attending a play.

Menzies would certainly have been at the other end of the financial scale and must have struggled to meet Edinburgh's high cost of living. One student, Thomas Ismay, the son of a Yorkshire clergyman, who paid just £10 a quarter, thought that this was the least costly accommodation to be found in the town: 'Twice a day if I chuse; have a hot dinner and supper. The dinner generally consists of a large Tirene [sic] of Soup, which I like extremely well, a dish of boiled meat, and another of roast. The mutton and beef is very good. Veal I have not seen yet; puddings, only one. Generally no supper, fish, eggs, beefstakes [sic] or what you please. Candles what you have occasion for, and a good fire.'

Menzies would not only have met Americans at university. Students came to Edinburgh from all over Europe, as well as from Ireland and many from south of the Border. What they all shared was less a common language, similar background, or comfortable financial situation, than, to a man, an initial dread of watching surgery and various treatments on patients. Charles Darwin, who stumbled out of an amputation being performed on a child a few years later and abandoned medicine for ever, was far from unusual. Others found that they had to leave the operating rooms before the bloody finale. At least for Menzies there was one comfort: botany classes were part of the curriculum, and fewer attended those than any other discipline. At the third year of study, only 17 per cent of students attended these classes. Few saw them as essential to their future career, knowing that herbalists were available in most town and country districts. What was the point of learning more when one could purchase a remedy close by? But for a potential naval physician, botany often turned out to be of huge value. Thousands of miles from home on foreign shores, the herbal supplies in his physician's boxes depleted or rotten, it might be his sharp eyes which picked out the herbs to replenish the stores, especially plucking up likely sources for counteracting scurvy. For many of the plant collectors who studied medicine, like Menzies, serious botanical study might just give them an edge in their careers, pushing open a door when a post called for a man who could be relied upon to spot the relevant herb in an unknown and alien landscape. But it was a marginal advantage.

By 1781 he qualified but did not immediately take the usual career path south, one that would almost certainly require taking to the road or a flea-ridden, rude berth on a small cargo ship to London. Instead, he went to Wales. For a brief time, he became an assistant to a surgeon in a practice at Caernarfon, Wales, but it must have been a fairly dull life, as he applied for a post in the British Navy. This, as he well knew, was the time-honoured solution for lads from less well-connected backgrounds who wanted to see the world and advance in life. Almost immediately he was in the thick of war, and, fortunately, on the winning side. In the long-standing wars with France, he was appointed a junior surgeon on board the *Nonsuch*, under Captain Truscott.

Dispatched to the West Indies to protect the sugar islands from takeover by the French, it was to be a bloody introduction to sea warfare. Sailing in formation under the command of Admiral Rodney, he observed at first hand the defeat near Dominica of the French admiral Comte de Grasse on 12 April 1782. The British Navy had been trying for some time to pin down this adopted hero of the American revolutionary forces, and the brilliance of the battle tactics by the winning Rodney won him elevation to Admiral of the Fleet. Descriptions of the carnage within both British and French ships were extreme even by the standards of the time.

While this was a brief respite for the British, whose grip on the 'colonies' of America was slipping badly, this battle also took home honours, raised morale, and, with seven French ships captured, was a welcome boost for a nation which felt its naval dominance bleeding away. As he amputated limbs, patched up appalling injuries from flying shards of wood, and watched as the steamy heat accelerated disease in badly mutilated sailors, Menzies must have tucked all this experience away and perhaps conceived of solace within the healing world of botany.

Quieter periods followed, with a voyage up the east coast of the United States, and he landed for some months in the chillier climate of Halifax, Nova Scotia. Perhaps he was just kicking his heels, or perhaps his early interest in plants was now firmly rekindled, but he found time to observe, note and collect plants from the east coast.

Menzies realised that his best bet was to ally himself with Sir Joseph Banks, the key figure in promoting the interests of fledgling botanists such as himself. He was effusive in correspondence in order to convince Banks that his main interest in life was botanising, sprinkling his letters with offers to search out species for Banks' collections of plants. Menzies was now demonstrating his considerable ability to be astute while pursuing with dogged determination a career which combined botany and medicine at sea, a combination which he hoped would catapult him further than a career confined to the vagaries of a naval appointment. With clarity, he deduced that his future lay in cultivating excellent relations with the important men in the country. Although but a lowly surgeon and possessing little in the way of personal influence, his insistence on keeping his name to the fore by constant letter-writing was his best bet. Menzies hoped that he would be seen as somebody who continued to help build Banks' reputation without presenting a threat. Menzies rarely demonstrated ruthlessness, but was terrier-like in his ambition. Moreover, he was careful not only to keep up a correspondence with Banks, but also to address many letters to his old mentor Dr John Hope at the Royal Botanic Garden in Edinburgh. Carefully he dispatched seeds and specimens of plants to both men. To Mr Hope he enquired always about Mrs Hope and the rest of the family.

He was hopeful of revisiting the Bahamas, but as it was April, the end of the dry season, few seeds were to be had, and with no other opportunities to sail back to Canada or visit the Bahamas again on board a naval ship, he returned to England and lost no time pursuing another chance, which this time might be to sail round the world. However the ships, the *Prince of Wales* and the *Princess Royal*, upon which he had heard about a berth he was determined to secure as surgeon, posed quite a problem. Understandably for a trading ship, no one on board was meant, on landing, to barter or gather anything for his own gain. Crew were often banned from trading on foreign shores on their own account, as the object of these merchant ships was to focus on the trade upon which they were engaged. But for Menzies, as a plant-seeking botanist, the accepted method of encouraging local people to assist him locate a plant in a short time was to offer presents in exchange. It was a very different approach from bartering for his own profit. Immediately Menzies directed an appeal to Banks.

> What I most regret is that we are not allowed – as the chief object of going sailing was to trade in furs – to barter or trade for any curiosities. I hope however we are not debarred from picking them up when they come our way. May I request the favour of a recommendatory letter to Mr Etches this at number 69 Watson Street, if he is known to you, as it may in some measure exempt me from this restriction; especially while my aims in collecting seeds, specimens and other curiosities do not interfere with the object of the voyage or in the interest of the company.

Menzies secured a minor triumph. Within days Mr Etches had a prompt letter from Banks, and replied,

> I was duly honoured with your kind favour by Mr Menzies, to which I feel myself bound to pay it every possible attention. I believe you are fully acquainted with the restrictions laid down in the articles of the former ships, in a young undertaking and of such

a nature as the present I presume such restrictions are absolutely necessary, but in the present instance it is my full intention to dispense with them in the case of Mr Menzies, so far as can have any tendency to be beneficial to science in general. I highly approve of his conduct and manners, as my younger brother 'who is part proprietor or' is going on the voyage. I gave him orders to pay every attention to Mr Menzies and to give him ample latitude in his pursuits and I have no doubt on his return he will confess having experienced the liberality which all recommendation, Sir, most certainly demands from me.

Etches was not slow to see another business opportunity:

A gentleman has made me a handsome offer to give him a passage in the *Prince* [*sic*] *of Wales* and to land him and a servant from one of the ships at the island of Otaheite, where he promises to reside for two or three years, till an opportunity may offer for his return. As we could accommodate another or two in a very handsome manner, if you know of any gentleman who would like such an excursion I shall be happy to treat with them and I would engage to fetch them back in the year 1788 or 1789, whichever would be most agreeable.

Menzies departed as wished, and on 14 November 1786 reached the Cape Verde Islands, and Menzies made sure that in his letter thanking Banks he mentioned that Mr Etches, acting on Banks' intervention, had made sure that Menzies had everything he needed for the voyage. The power of Banks was not to be underestimated. Before departure from Yarmouth Menzies had visited Banks' herbarium which had been collected at King George's Sound, and most of it was new to him. In his letter to Banks, Menzies also mentioned that before leaving London he had visited Kew where 'Mr Aiton [the head gardener] received me with all his usual hospitality and friendship . . . [and] *Hedyotis caerulea* [creeping bluets, a rock plant with star-shaped violet flowers] which he said he raised from some seeds I sent from Nova Scotia . . . I shall use every means in my power to send you *Wintera aromatica* [probably wintergreen, another ground-cover plant with pale to deep pink flowers] from Staten Island.' At the end of January the *Prince of Wales* anchored at New York Harbour, Staten Island, and she was to stay there for two weeks. The weather was appalling but this did not prevent Menzies and the ship's captain Colnett from exploring. The latter wrote in his journal that they walked through rushes 'such as I have seen in Marsh grounds in England'. They shot hawks and their dog found a seal which was holed up in a rock. The latter they transported back to the ship for food, preserving the skin. As promised, Menzies did find the winter-flowering *Wintera aromatica*, dispatching it to Dr Hope and Banks. Then the voyage continued. The ship rounded Cape Horn, and in early July when north of present-day California, the captain was able to report that even the members of the crew who were suffering from scurvy were cheered by seeing the trees on landing. They traded for furs around Nootka Sound, near present-day Vancouver, and then spent the worst of the winter months in Hawaii.

In the spring they returned to Nootka to trade for more skins but with moderate success. The ships then sailed on to Canton in China, and finally back to England, landing on 14 July 1789 at the Isle of Wight, from where Menzies reported immediately back to Banks.

I have the pleasure to acquaint you of our safe arrival thus far in the British Channel after a tedious voyage round the globe of nearly three years during which time the officers and crew of both vessels continued pretty healthy, notwithstanding the many and sudden vicissitudes of climate they went through in that period . . . excepting once which as soon as we crossed the Equator in the Pacific, in going northward most of the men and officers were laid up by an inveterate scurvy of which however they all recovered soon after our arrival at Nootka.

We have lost only one man during the whole voyage, who died 20 days ago . . . On the west side of North America in a remote corner inland I saw the natives have a short warlike weapon of solid brass – somewhat in the shape of the New Zealand weapon (like a truncheon in shape) about 15 inches long; it had a short handle with a round knob at the end and the blade was of an oval form, thick in the middle but becoming gradually thinner towards the edges and embellished on one side with an Escutcheon inscribed Joseph Banks Esq. The natives put a very high value on it, for they would not part with it for considerable offers. The embellishment work is nearly worn off by their great attention to keeping it clean. On which I beg leave to remark that the inscription except on such durable tokens left among savages ought to be deeply impressed. Announce the date when, and if possible the place was, where it was Left . . . for it remains to be determined through what intermediate conveyance the above instrument reached that place as I am almost certain we were the first Europeans with whom the natives ever had any direct communication.

To commemorate this discovery I have given your name to a cluster of islands round where we were then at anchor and in the course of a few days I hope I shall have the honour of pointing out to you their situation and extent on the Chart I have made of the coast – as also of presenting you with a memento from that and other parts of it.

Menzies divided up his seeds and plants, sending them both to Banks and up to Edinburgh. By an amazing coincidence, Menzies had stumbled across the truncheon which Banks had either left behind or given to the local people on his circumnavigation of the globe with Captain James Cook on the *Endeavour* in 1770. Dr Hope had died two years earlier, but his successor at the Botanic Gardens, Rutherford, received the following letter:

Though I have not the honour of being personally known to you yet I have taken the liberty of sending you, a small parcel of seeds, hoping they will make some additions to that valuable collection of the Botanic Garden, and thereby prove acceptable. They will arrive on Tuesday night by coach at Robertson's Black Bull at the top of Leith Walk. These seeds were collected by me in the late voyage round the world . . . [some] were collected on an island near Cape Horn in January and February 1787 and those marked from north America were collected on the West Coast north America between the latitude of 48° and 61° north in the summers of 1787 and 1788 so both will stand at the open air in Edinburgh. The rest were collected in September 1788 at the Sandwich Islands [Hawaii] and on February 1789 at the south end of Sumatra; these must be considered as houseplants.

He continued to tell Rutherford how much he respected Hope and appreciated all that had been done for him. He also told him of his appreciation of being able to access the books and herbarium belonging to Banks, apologising for gaps in his (Menzies') knowledge and explaining that he had few botanical reference works on board. Just in case Rutherford had still not made the connection, he also pointed out that a gardener working under him was Menzies' brother William. William had, however, been in London with him for the last ten days. Ever the supportive brother, Menzies was also quick to reassure Rutherford that William and himself had wasted no time in London and had already visited many of the great plant collections. For example: 'on Saturday last I introduced him to Dr Pitcairn who was kind enough to offer him any plant that could be spared out of his collection at Islington which is very considerable. Mr Aiton at Kew has also promised to give him some rare plants – he was with me here today when my brother received your letter and made his promise to be with him at Kew tomorrow to re-examine that great and unparalleled collection of rare and scarce plants.'

No sooner had Menzies unpacked his specimens than he was preparing for a very prestigious voyage aboard the aptly named *Discovery* under the command of Captain Vancouver.

George Vancouver was born on 22 June 1758, at King's Lynn, the sixth child of Bridget and John Vancouver, carrying a name that was a corruption of van Coevorden. An ancestor of George had married an Englishwoman called Sarah, and George's mother and therefore himself and his five older siblings were descended from Sir Richard Grenville, a naval hero. Growing up in King's Lynn, his childhood was enmeshed with the sea. His father held the fairly important job of Deputy Collector of Customs and Collector of the Town Dues. King's Lynn was a thriving port with ships from all over Europe clogging the docks; there was an occasional fracas as naval press gangs roamed the streets looking for likely recruits that they could force on board. George, however, followed a more advantageous route to the sea.

His first taste was aboard the *Resolution*, under Captain Cook, who set off on his second voyage of discovery to search out the mysterious continent which he was certain still eluded him in the southern latitudes. Cook was backed in this certainty by Sir Joseph Banks, who had accompanied him as botanist on his first circumnavigation. Vancouver was therefore in the midst of the confusion and fighting on 13 February 1779, when Captain Cook was stabbed to death in Kealakekua Bay, the Sandwich Islands (now Hawaii). Vancouver was clubbed by an oar, but apart from the indignity, he was lucky to suffer only the loss of his hat. Eleven years later, he was appointed Commander of the *Discovery*, and a second ship, the *Chatham*, with a mission to explore the north-west coast of North America, taking in parts of what are now America, Canada and Alaska.

Menzies had been appointed as a surgeon on board, as well as official botanist. If he had gleaned much about Vancouver from naval chatter he kept his counsel regarding his feeling on his captain. Sir Joseph Banks, however, keen as ever for new natural history finds, with his ear always to the ground for news that might bring him new species, collections and exciting discoveries, had grave doubts. Already he had picked up rumours about George Vancouver, and they were not promising.

Banks had insisted that the ship should have glass plant-cases on board to house all the expected new finds. Vancouver was horrified that Banks would impose his wishes on his ship. To insist on a botanist travelling was just about acceptable, but with the added command that the ship would be required to have glass plant-cases on board was one step too far. Vancouver objected. Sir Joseph Banks, one of the most important men of the era, was outraged. He won the argument and Menzies and the container boxes went on board. Banks lost no time in voicing his displeasure and wrote to Menzies that how 'Captain Vancouver would behave to you is more than I can guess, unless I was to judge by his conduct to me – which was not such as I am used to receive from persons in his situation . . . As it would be highly impudent in him to throw any obstacles in the way of your duty, I trust he will have too much good sense to obstruct it.' He advised Menzies to note down in his journal any incidents of Vancouver obstructing him, which could be used in evidence upon his return.

It is just as well Vancouver was ignorant of these missives. However, his troubles were just about to begin. Among the queue of aristocratic young gentlemen who were being 'recommended' to him, which meant he was required to take them, and in no way could he refuse those who were his superiors in class and career, was the Honourable Thomas Pitt. Pitt's father, Baron Camelford, was the cousin of Prime Minister William Pitt, and his sister was about to wed the First Lord of the Admiralty. Thomas Pitt was a sixteen-year-old troublemaker. Tall, well built and arrogant, he was also already the veteran of a previous hair-raising voyage. He was on board the HMS *Guardian* when, loaded with convicts, the ship's collision with an iceberg holed her so badly that it was only thanks to the highly skilled hands of her captain, Lieutenant Edward Riou, and the good fortune of barrels rolling in to plug the gap, that she was able to be sailed at all. Despite being accompanied by young Pitt on this 2,000-mile journey before the ship beached at the Cape, Vancouver took the unusual step of refusing to grant Pitt a signed certificate of service, the vital 'reference' which most seamen required to gain employment on another ship and advance their careers. Pitt was to haunt Vancouver. Also on board was a Hawaiian, Towereroo, nicknamed 'John Ingram', who had been given a passage home, but who wished to visit the north-west coast of the Americas before leaving the ship.

The voyage was planned to go by way of the Cape, New Zealand, Tahiti and Hawaii, to Nootka Sound, California, the Galapagos Islands, Chile and round the Horn at the tip of South America.

The European world stage which they were leaving behind was calmer in the sense that England was no longer under threat from across the Channel, but as for Europe itself, it was an uneasy calm. The newborn United States of America was a *fait accompli* and Catherine the Great was busy invading Poland and dividing up the spoils. For once the seas around Britain were more or less devoid of danger from their European and American cousins, which ensured a relatively safe passage across the Atlantic – if they disregarded stalking pirate ships – but on land the forces were gathering. If Banks, Vancouver and Menzies could have been swallows over Europe, it would have appeared an excellent moment to sail off in the opposite direction, away from the turmoil. The French were turning themselves inland with fighting intent, and were on the threshold of declaring war on Austria and Prussia. The Reign of Terror was just a couple of years away.

Despite the *Discovery* having never been to sea before, or even subjected to sea trial, overloaded with young 'gentlemen', and accompanied by the wallowing *Chatham*, she made for the open sea. But bad luck hit the voyage even before leaving home waters. At Falmouth, they arrived just as the press-gang men were in full cry. For the other men on board Vancouver's ship, the prospect of war, and therefore fat prize money for captured ships, was enticing, and few would pass up this opportunity. Some of Vancouver's men immediately deserted, determined to take part in the impending war and enjoy the spoils. Vancouver had to replace deserters and many of the men he signed up he eventually discovered were sick and unfit. When they finally set sail, on 1 April 1791, Vancouver must have been thankful to leave the shores of Britain behind. He was being dispatched not only on a voyage of discovery to chart the coastline of the west coast of the Americas, but also to deal diplomatically with Spanish claims to the area. He was thirty-two years old and few of his seamen were any older. This collection of young men, packed into two smallish ships, was to spend four and a half years in frequently uncharted waters, with the possibility of attack from ashore, pirates and ships from countries who might be at war with Britain. Many ships' commanders of the previous few hundred years had been in this position before, and their command usually rested on absolute power, and fierce and frequent brutality.

Vancouver's character is almost impossible to pigeonhole. On the one hand he was ambitious and capable, a skilled diplomat when abroad, courageous and a skilled seaman, with an unexpected streak of compassion. On the other, he appears to have had a blind spot when it came to dealing with the powerful men who controlled his future, and to maintaining the morale of his crew and welding them into a loyal team. Added to this was his emerging ill health.

Before the expedition had reached Tenerife, he was ordering the decks to be smoked clean using gunpowder and vinegar, which even he realised was 'inconvenient and disagreeable', as he recorded in his journal. At Tenerife, the ship's crew, reluctant to be ordered back on board after a night in the dockland bars, were involved in a running fight, and Vancouver himself was toppled into the water. By August they had reached South Africa, when his chosen surgeon, Cranstoun, became so ill that he was forced into an about-turn and appointed Menzies in his place, a role he had denied him but a few months before. As they sailed towards New Holland (now Australia), the *Chatham* fired a distress signal when one of her stern windows was blown in allowing a dangerous amount of water to flow in rapidly, but the *Discovery* sailed on oblivious. Although managing with difficulty to rescue the situation alone, the commander of the *Chatham*, Broughton, was not pleased.

As they arrived at the south-western corner of Australia at what is now Albany, Vancouver named the bay King George III's Sound (now King George Sound). Here twelve days were spent at anchor, the sick being put ashore to recuperate, and Menzies could at last gather specimens. Having charted this area, the ships sailed on south round the tip of South Island, New Zealand, and set a course for the Sandwich Islands (Hawaii), calling in at Matavai Bay, on Tahiti. They always set a rendezvous point in case they lost sight of one another, and this happened frequently.

The *Chatham* was always deemed to be the slower ship, often losing sight of *Discovery*. While this was fact, and Vancouver was well aware of the difference between the two vessels, it nevertheless added to his ill temper. By the time the ships reached Tahiti Vancouver was far from relaxed. His last visit to the area had ended in dispute with the local people, and the murder of Captain Cook. He appeared worried that the same scenario might present itself, and his short temper was fraying badly. To add to his ill feelings, there had been a theft of an iron axe and a quantity of clothing. Then, to cap it all, Towereroo jumped ship and took to the hills with the local chief's daughter. Vancouver reacted at once, concerned that Towereroo, with his knowledge of firearms, might change the balance of power in the area. He threatened to burn down the chief 's house in order to assert his authority, and for good measure he put a rope round the neck of a man whom he suspected was the native thief, and finally threatened to burn down the whole village and the natives' boats. Towereroo was promptly found and returned to the ship, but the rest of the ship's company, who had been dreaming of the delights of the island women, were denied time ashore by Vancouver unless on duty.

Pitt had already been caned in the midshipmen's mess for various misdemeanours; now he made an unsanctioned sortie ashore producing a haul of six stolen pigs. Vancouver requisitioned them and ordered their slaughter followed by their consumption by the entire ship's crew in lieu of one day's rations. As purser as well as captain he was in control of finances and this would have made a useful credit balance.

One month later, with Vancouver anxious and on edge, quick to quash any behaviour which questioned his authority, they set sail for Hawaii, arriving exactly thirteen years to the month since Vancouver had been on the shore with Captain Cook and witnessed his murder. To compound Vancouver's nervousness, they anchored at Kealakekua Bay, the actual scene of the murder, and, whilst out walking on the beach with Menzies and Johnstone, Vancouver reacted almost hysterically to a collection of fires burning up on the hillsides. In vain did Menzies try and convince him that these were only in order to clear dead grass. Vancouver was certain they were a preparation for attack, especially as the natives were now well supplied with guns and had, only a couple of years before, attacked an American schooner, killing all on board.

Vancouver ordered a couple of midshipmen to seize a native canoe and take him back out to the ship. They conveyed him safely offshore away from the heavy surf of the beach, but the inevitable happened. Unused to the canoe, the young men capsized with Vancouver aboard. Vancouver swam on alone to the safety of the ship, while the two non-swimming midshipmen clung to the upturned canoe. His wrath was turned on the hapless youngsters clinging to the canoe, whom he accused of trying to kill him by deliberately capsizing the canoe. As the voyage wore on, his anger turned on others in the crew. The ship's carpenter, who by this stage in the voyage was probably exhausted from his efforts to keep the masts in order, was confined to his cabin for answering back when Vancouver criticised his workmanship. Menzies records little in his journal of the voyage concerning Vancouver's moods at this stage. This was to emerge much later. Most probably he was by now used to Vancouver's mercurial temper and well aware of the strain such a long sea voyage would put on the crew. He concentrated on filling in his journal with details of what he could see of the coast by telescope, as they passed what is now Oregon.

However, if morale was low amongst the ship's company, at least Vancouver felt that he was now approaching the area for which he had been dispatched just over a year previously. Now was his chance to complete the surveying of the coastline, his chance to make a dent in the history books, and at least they were away from the place where Cook and indeed he himself had come under attack. As they sailed up the coast from around thirty-nine degrees north, to around a hundred and fifteen miles north of San Francisco, Vancouver busied himself thinking up names for every point they passed. For Menzies, their proximity to the land must have been tantalising:

> This coast preserves nearly a northerly direction and affords in many places particularly to the Southward, most beautiful prospects of hills and valleys varied with wards and pastures mounting up their sides, presenting to the eye delightful rural landscapes, and to the mind the idea of a mountainous country in a state of high cultivation, which I could not pass without often regretting my not being able to land and examine it more particularly . . . It is abounding with extended lawns and rich pastures, not inferior in beauty of prospect to the most admired parks in England.

Ashore lay a forest of which he might have heard tales, but was now possibly just able to make out with his telescope. These were the giant redwoods of what is now named Redwood Empire. Giant redwoods stretched for miles. Further north he and the midshipman Thomas Manby climbed up the masthead and deduced that the muddy water at the entrance of what was called Cape Disappointment on their charts might not be quite so disappointing as it sounded; they were correct. The muddy inlet, shrouded in fog, disguised the entrance to the mighty Columbia River. By the time the ships dropped anchor in the sheltered inlets just to the north-east of what is now Olympic National Park, Washington State, and south-east of Vancouver Island, on 2 May 1792, Vancouver had reason to feel fairly satisfied, even to the extent of giving his men their first day's holiday since leaving the South African Cape almost eleven months before. But this was only after they had beached the ships and given them a thorough overhaul.

Their most astonishing surprise so far had been to encounter a sailing ship, the *Columbia*, nine months out from Boston, whose captain, Robert Gray, was reputed to have been the man who had managed to sail round the fabled and elusive North-west Passage, from the North Atlantic, round the edge of Canada, round Alaska and down into the North Pacific. This was the longed-for alternative to the torturous passage through the South Atlantic, round the dreaded Cape Horn and up the Pacific to the north-west American coast. Gray admitted straight away to Vancouver that he had never in fact managed this trip, merely venturing up the

Straits of Juan de Fuca for fifty miles. There, upon finding that the sea otter pelts he required were scarce, and that all but a few traders had given up the whole idea of bringing pelts to the shore for barter and had left the area, he turned back. Nevertheless, Vancouver had sent Menzies and Puget to secure any information Gray may have possessed regarding the north-west coastal area, in order to fill in a lack of information on their charts.

The search for sea inlets and suitable sheltered harbours was fraught for all voyagers. If they passed by at night, or drifted offshore in fog, they frequently missed many of the possible harbours. Although Gray had warned them to be wary of the local tribesmen, some came to trade with the ship, bringing fish and venison, and these were accepted with alacrity and pronounced extremely good. But misunderstandings swiftly followed. The same locals brought along two children, aged six or seven, and wanted to exchange them for copper sheeting or muskets. Although the ship's crew were horrified to a man, and refused the offer, they were not to know that Spanish ships had regularly taken the children believing that they were rescuing them from slavery or cannibalism. Later, when cooked venison was offered to the local tribesmen, the latter refused with horror, imagining it to be human flesh. On 7 May, accompanied by Vancouver, Menzies set off to explore an inlet they later named Hood Canal. Other members of the crew set off in differing directions to survey, as they could get closer ashore in small boats allowing more accurate surveying.

At last Menzies could accumulate some botanical specimens, and a satisfying peace descended. Even Vancouver appeared relaxed and wrote that, 'The serenity of the climate, the innumerable pleasing landscapes, and the abundant fertility that unassisted nature puts forth, require only to be enriched by the industry of man with villages, mansions, cottages, and other buildings to render it the most lovely country that can be imagined.' Anxious to chart the most reliable maps of the area, with an eye on his future, and still in his mid thirties and with most of his health still intact, he took the decision to use the small open boats available to them to carry out this task. For Vancouver it was without doubt the correct decision. However, his sweating sailors, rowing relentlessly every day for hours at a time, felt closer to being galley slaves. The local people came willingly to meet them ashore, bringing offers of what they described as hyacinth bulbs, but were in fact *Camassia quamash* (common camas). But what excited Vancouver was the wealth of timber, which he – correctly – estimated to be excellent for shipbuilding for the British Navy.

Menzies and the other officers from the two ships appeared to be impressed by the local landscape, the 'spacious meadows' elegantly adorned with clumps of trees and the acquiescent and friendly local peoples. They were invited ashore and indeed into the huts by the shore, and although their refusal appeared to mortify the local women, no ill feeling ensued. Enough understanding took place for the locals to ask Whidbey, a warrant officer on board employed in surveying, why his face was painted white. Whidbey answered by unbuttoning his waistcoat and exposing his matching pale chest, to much general amazement. Food was brought out to the boats in return for trinkets such as beads and tobacco, and relations became so cordial that the locals enjoyed short sailing excursions. Being June, even the weather was moderately calm.

Imbued with satisfaction, Vancouver, accompanied by gunnery salutes from the ships, stood ashore at what was named Possession Point, and claimed a chunk of land in the name of Great Britain close to the area around present-day Vancouver. That some of the land had already been claimed by Spain was ignored, as indeed was the fact that he, Vancouver, was acting entirely on his own volition. Nothing in his instructions from His Majesty's Government permitted him to take this step. Later, the North West Company and the Hudson's Bay Company would regard themselves, and the tiny minority of Europeans within the vastness of the north-west of Canada, as akin to feudal overlords, quasi-kings in their own mini-fiefdoms. Vancouver was leading the way.

His bonhomie was short-lived. Broughton, on board the *Chatham* which was gripped by strong currents, lost his irreplaceable stream anchor. Vancouver's temper erupted. A short time later Broughton requested more yellow- or clear-coloured varnish to complete the painting of his ship, but was refused. He was forced to paint the other side of his ship black, a punishing and possibly vindictive move by Vancouver that publicly humiliated Broughton. So with one side pale and the other black, his error in losing the stream anchor was visible to all for months to come. If Menzies was concerned about the volatility of Vancouver's mood, he kept his counsel. On the first of the *Endeavour's* circumnavigations of the globe with Cook, Banks was the diplomat who had smoothed over this type of incident, although he never appeared to interfere with the running of the ship. Banks was an expert and reassuring pourer of oil on troubled waters. Vancouver had no such influence on board.

Less volatile characters than Vancouver might have felt it prudent to hold their tongues for a few days. After all, Vancouver was not only demonstrating his displeasure at what might well have been bad seamanship, he was alienating and undermining the authority of his brother captain. Far from home, and with no other vessels in the vicinity, it was essential that the two men should enjoy mutual support. No one knew what the future might hold. Broughton was in command of the only vessel likely to support or indeed rescue Vancouver if he was in difficulties. Broughton was his second in command, the first person to whom he could turn for discussion, for his opinions, and for companionship. Not only had Vancouver risked souring this relationship, he had also taken the most effective means possible to humiliate the commander of the *Chatham* in front of his crew.

The next in line to feel Vancouver's wrath was Thomas Manby, the master's mate, who had been put in charge of the pinnacle. Leaving in tandem with Vancouver's small boat, the two had then deliberately separated, in order to chart and explore different inlets. When Manby had returned by the same route to the point at which the two had gone their different ways, there was no trace of Vancouver, who had decided to take a different route back to the ship. Without a compass he and his crew spent three days lost and without provisions. They gathered mussels that gave them all serious food poisoning, and, weakened and debilitated, they finally crawled back to the anchored ships and the wrath of Vancouver. Manby wrote in his diary that he found it difficult either to forgive or forget such a verbal assault. But Vancouver's wrath and disappointment was not so much directed at Manby, who just happened to be a convenient whipping post. He had espied, to his horror, a pair of Spanish ships in the vicinity. Worse, they had already ventured much further than Vancouver.

He was faced with a predicament. Not only might his personal future as a successful explorer be in jeopardy, he also had little idea of what had happened on the world stage for two years. So he was running a considerable risk in falling in with foreigners, with whom, on the other side of the world, Britain might well be at war. Ever the pragmatist, he realised that both parties, the British and the Spaniards, were in a remote, wild and largely uncharted area and might need to rely on one another if in difficulty. Building up a rapport was a good insurance policy and worth the risk. On hearing that they had explored further afield than his own expedition, he realised that perhaps his only chance, and certainly his best bet, was to join them. Vancouver dredged up his diplomatic skills. He knew full well that the Spaniards might well have already charted areas he was all set to explore, or worse claimed them for Spain. He did not know if perhaps they had already found the North-west Passage, and were keeping this information to themselves. If either of these possibilities came true, then he would have to feel that his entire expedition had failed. He sensed that his back might be to the wall, and he was confident he could rise to the occasion.

Cunningly, but with easy agreement on the part of the Spanish, he insisted that the eastern edges would be covered exhaustively by his own men. Should the much sought after North-west Passage, in the shape of a river that would bring them out on the eastern shore of the United States, exist, then he would claim the glory and for Great Britain the exclusivity of the passage.

Together he and the two Spanish captains, Galiano, commander of the *Sutil*, and Valdes of the *Mexicana*, carved up the areas for each to explore. In theory this would keep both sides happy. Mostly it was successful, although as Puget and Menzies doggedly rowed up the Toba Inlet they met the Spanish captain, Valdes, exiting this area. Valdes helpfully confirmed that this was indeed an inlet and not worthy of further investigation. Menzies and Puget listened, but fearful of disobeying Vancouver's orders, carried on regardless. Probably, Menzies and Puget either did not dare to come back without carrying out Vancouver's orders to the letter, or they realised that Vancouver had a definite agenda regarding the search for every available inlet which might lead to the elusive Northwest Passage. Valdes, however, watching Menzies and Puget continue on, disregarding his advice, immediately complained to Vancouver that he found this attitude less than trusting. Thankfully for Menzies and Puget, Vancouver responded that he himself had been expressly ordered to see all the inlets himself. Less than thankfully, Menzies and his companions landed at a deserted village, which on exploration was found to be alive with vermin and fleas. In a desperate attempt to rid themselves of the pesky fleas, they dashed back into the water, stripping themselves of their clothes, which they submerged in an attempt to drown the offensive fleas. Even this did not work. Plan two was to chuck their clothes overboard, tying them to the boat and dragging them behind. Plan three was to put their clothes overnight in boiling water. This appeared to do the trick.

The days continued with hours of stiff rowing in fractious tidal conditions in small boats as they probed miles of coast. The *Discovery* grounded on submerged reefs and heeled almost over. Amid much anxiety she rose up and crashed down again on the ebbing tide. After finally floating off, it was the *Chatham*'s turn to sail on, which she did without mishap. Between all these tense moments, there was time for relaxation and comfort, away from the confines of the ship. The officers of the two British ships enjoyed hospitality of a high order from Spaniard Don Quadra, who was the Governor of St Blas and Commander in Chief of the Royal Navy of Mexico and California. Menzies was most impressed by the construction of his residence which included a two-storey house, dining facilities with silverware, fresh rolls served every morning, and vegetables from the established and flourishing gardens. They also gained from the good relationships the Spaniards had made with the Native American chief, Maquinna. The commander, Captain Quadra, invited the British officer and hosted with what must have appeared in such a remote place lavish hospitality:

> He lived onshore at a very decent planked house, in a very sumptuous manner considering the situation where he kept an open table, I may say, for the officers of every vessel that visited the port, and supplied them on board with vegetables milk and other refreshments . . . Don Quadra, on our arrival at Nootka had put himself [interjections] under my care for a severe headache, of which he said he complained upwards of two years, and I was extremely happy that my endeavours was [sic] in some measure serviceable towards his recovery.

Menzies noted that the climate was:

> Exceeding favourable in so high a latitude, and the soil, tho' in general light, would I am confident yield good and early crops of most of the European grains; its being so intersected with branches of the sea affords easy and commodious communications and its short distance from, and free access to the ocean by De Fuca's Straits may likewise in time be much in its favour, for its most Eastern boundary is not above 50 leagues inland and is formed by a ridge of mountains in many places covered with perpetual snow . . . there is a greater variety of hard wood scattered along the shores here than I have observed on any other part of the coast.

He became increasingly excited about new and rare plants, some of which, he felt, could not be positively identified until he saw Banks.

But of course, the fur traders merely saw European settlement as impinging on their trade. The fewer the habitations, the less the potential destruction of otter and beaver homelands. Menzies was sensitive to this. Inland, he concluded, was far less rich in fur trapping and hunting, perhaps owing to the density of the forests and steepness of the terrain. Conversely, it was a rich picking ground for plant life and especially timber.

The northern tips of Queen Charlotte Island produced more furs for the Chinese trade than all the rest of the coast together, he wrote to Banks, adding with considerable astonishment and foresight that the area required immediate regulation, as 'no less than thirty vessels have been on the North West Coast this summer, comprising English – twelve, Spanish – six, American – seven, Portuguese – four and one French'. As for the plants, he found several species of penstemons, rhododendrons, *Lonicera*, *Claytonia*, *Arbutus*, *Polygonum* and *Spireas*, all of which many years later became common garden plants in Europe.

Vancouver was now faced with a lack of sure direction in his dealings over the land claims in the area, being made by Spain and Britain. He also had to reshuffle his crews owing to the murders of the captain and a crew member on board the *Daedalus*, the ship which had been en route to join them, bringing stores and relieving some of the crew. Vancouver appeared to feel the weight of history on his shoulders, and did not want to make a wrong decision in what he recognised was a matter of major international importance. He was just a sailor, and, while diplomacy on board the ship within ordinary life was well within his grasp, in international matters where his country, not to mention his career prospects, was on the line, he needed to search out a second opinion. The death of his friend, the captain of the *Daedalus*, removed someone with whom he felt on an equal footing, and who he could trust as a friend. He compensated by imposing his power through shuffling the crews around.

It was now that a simmering animosity between Vancouver and Menzies came to the surface. Vancouver wished to replace the invalid surgeon Cranstoun with Menzies who, although being well qualified and indeed having originally sought the job, now demurred. Initially Menzies would have jumped at the chance, but having been appointed as the naturalist on board, he realised that he could not do both jobs well. There was a brief stand-off. Vancouver did not possess the necessary authority to overrule the Admiralty and appoint Menzies to the post. Equally, Menzies was reluctant. It was only when Vancouver brought things to a head by demanding that Menzies refuse the offer in writing that the matter was resolved. Menzies accepted the offer, noting in his diary that his decision was based purely on nervousness that a refusal might jeopardise his naval career. He also wrote in his account for Banks that Vancouver had promised that his new duties would interfere as little as possible with his botanical pursuits, and Menzies, ever the pragmatist, thought that the advantage of having another cabin, and a couple of assistants, outweighed the disadvantages. Also in his favour was that he observed the ship to be generally healthy so he foresaw little in the way of an imminent, huge workload. And finally, for good measure, he thought that his added duties as surgeon were in reality little more work, as he had been filling in for the ill Cranstoun anyway. In other words, there was little change. It was an optimistic assessment, and during the voyage home he acquitted himself well. He was to report later that during that section of the voyage when he was in charge, he lost but one man, and the dreaded scurvy was kept mainly at bay. This might or might not have been assisted by his special brew, a recipe which he had picked up at Nootka. It was a potent mixture of spruce needles and molasses, which was possibly antiscorbutic. He dosed the crew with it when necessary. Finally on 12 October 1794, they turned the ship southwards, leaving Nootka behind and heading for San Francisco and the long voyage home.

Arriving at the port of San Francisco one month later, they then proceeded on to Monterey, where the ever effusive Don Quadra had been waiting for six weeks to receive them. In his journal, Menzies was equally gushing:

> In short I cannot express to you the vast attention and civility which have been paid to us at these two settlements since we

came to California: every kind of refreshment which the country affords is supplied to us in abundance and without price, and not only for ourselves, but they have given us what cattle and sheep we choose to carry with us for the Sandwich Islands and our friends in Botany Bay, so that we carry with us to these islands four cows, two bulls and the same number of ewes and rams. The *Daedalus* sailed yesterday from hence for Botany Bay with twelve cows, six bulls and the same number of ewes and rams on board for that settlement. The amazing rapid increase of stock in this country is chiefly owing to the indefatigable industry of the worthy Fathers at the Missions who have constantly received us with a hearty welcome; they rear besides maize wheat, peas, beans, potatoes and all kinds of vegetables sufficient for their own consumption: but except a few apples they have not yet been able to raise any fruits. The cattle belonging to the two settlements and four adjacent missions are averaged at 2,400 heads, besides horses and vast flocks of sheep and goats: and if we except the natives the number of inhabitants men, women and children do not amount to 500.

Menzies was one of the first to send back seeds from the pine forests of the region, as well as seeds of plants that he thought would provide interest, but not necessarily be successful in the British climate. Unfortunately, since it was now November, plant and seed gathering was difficult. Deciduous plants denuded of leaves were impossible to identify, and the seeds were almost all scattered. He was also up against better equipped botanists.

Menzies remarked sorrowfully that the Spanish had two botanists in the area, both well financed, with an excellent draughtsman for their use, and were busy spending the Spanish king's gold on a horticultural magnum opus, *Flora Mexicana*. Even more galling was the fact that with a team working solely on this masterpiece, it would even be published before the *Discovery* hit the shores of England. Furthermore, the Spanish, under able navigator Señor Malaspini, were in command of a pair of ships purely employed to chart the coasts and the Philippine Islands, then down to Peru and Chile. Ruefully Menzies added that 'the Spanish mean now to shake off entirely that odium of secrecy and indolence with which they have been so long accused'.

Vancouver's expedition pottered southwards, touching on several points in California, having skirted the Islas Marias. Menzies was able to send some seeds back with Captain Broughton, who had been put ashore at St Blas on the Mexican coast and dispatched overland through Mexico around Christmas, 1795. From thence Broughton accompanied a party of Spaniards who all made their way overland to the Gulf of Mexico, and picked up a ship bound for Europe. Therefore, observed Menzies with more than a hint of longing, they would be in Europe in an almost unbelievably speedy four months. Destined to go round the Horn, it would be at least double that time before Menzies would see the coasts of home. Sailing south, they touched on the northern Galapagos Island whose appearance he did not find endearing, describing it as a 'dreary barren island with some high mountains, whose sides here and there are covered with tufts of dry shrivelled grass, and a few low scattering bushes while all the lower ground was strewed over [with] the black rugged lava and scoria [basaltic lava fragments ejected as fragments from a volcano], without the least appearance of soil, vegetation or fresh water; its shores inhabited by vast numbers of seals, penguins, guanas [iguana] and several other kinds of Lizards and snakes.' Thirty years later another Perthshire Scot, David Douglas, passed by and was frustrated by the short time he could spend ashore there, his curiosity sharply aroused by the strange animals around. Ten years after that Darwin landed and began his long journey of discovery into the true origins of species.

Vancouver and his crew had far more pressing thoughts. The mast had split and every island they called upon had little water. Menzies had also lost many of his tender plants. With one final landing in prospect, the voyage was sinking into a morass of bitterness and recriminations. By the end of March they were in Chile, where an encounter with the Irish-born Ambrosio O'Higgins (*c*.1720–1801), Govenor of Chile and father of Bernardo

O'Higgins, who became the first leader of the new Chilean state in 1817, led to a banquet and from thence to a singular puzzle. By the time he was entertaining the officers from the *Discovery* he was in his late seventies, having been variously a great campaigner, soldier, adventurer, romantic, social reformer and now President. While the British were banqueting with him, one puzzling nut on the dinner table led to a unsubstantiated tale which has been firmly attached to Menzies ever since. Supposedly, Menzies gazed curiously upon a dish of finger-shaped, bronze nuts, which were much enjoyed by the local hosts. He is supposed to have taken some and pocketed them. By this circuitous route, the *Araucaria araucana* or monkey puzzle tree arrived back in Britain to adorn Victorian parklands from the English Channel to the north of Scotland. Menzies did not mention pocketing the seeds, or indeed planting them up on board the *Discovery* and introducing them to the United Kingdom, but there has never been a credible tale of anyone else bringing them back. So Menzies' name remains attached to the introduction, and certainly the dates of introduction fit snugly with Menzies' visit to Chile.

As the ship set sail for home, plants, and their care, and incidents of discipline on board heralded the onset of a more sinister time aboard the *Discovery*. Relationships deteriorated, leading to a chain of events which even today is shrouded in innuendo and cover-ups. There is no certainty that the truth of much of the ship's final months ever emerged.

At some point, when north of the Equator and close to the island of St Helena in the South Atlantic, the ship was drenched by a heavy storm. Exposed on deck were the carefully nurtured plants gathered from the other side of the Americas by Menzies over the course of three years. In his role as botanist he had been allocated a servant to help him care for the fragile plants and seedlings, which were protected under cover on deck within specially built glass cases. On 28 July 1795, only three months short of home, the glass cases were left open, the manservant having neglected to cover them, and the plants were ruined. Menzies was horrified, aghast, and probably frightened at the prospect of losing the very plants upon which his salary and reputation depended. Then he was furious. He complained to Vancouver, who not only sided with the seaman, but put Menzies under arrest. Menzies detailed the occasion to Banks in a letter sent to him from Shannon in Ireland, explaining that he had never kept 'his' manservant from the necessary shipboard duties, and that the same manservant was an excellent worker, emphasising that he 'believed it will be allowed by the officers that he did more ship's duty before he was watched [put on watch] than any other idler on board'.

But the very fact that Vancouver had overridden Menzies' complaint that the seaman should have been caring for the plants and ignored Menzies' entreaties to chastise the seaman, instead ordering the man straight back on watch without any reprisal at all, finally squashed Menzies' authority.

Menzies had lost his dignity, his authority over his manservant, his precious plants, and his freedom. For three black months he sat alone in his cabin under tight arrest. He tried to write up his journals, but he found careful writing difficult with the ship's movement. He waited with trepidation to hear what Banks might say, and determined to throw himself on the mercy of the Lordships of the Admiralty, and place his future both in their hands, and in those of Banks. When Vancouver demanded his journals of the voyage, Menzies refused.

When the *Discovery* docked, Sir Joseph Banks backed Menzies against Vancouver. This was not auspicious for Vancouver. Furthermore, another young crew member, the adolescent, troublesome bully, the Honourable Thomas Pitt, second Lord Camelford, whom Vancouver had had regularly flogged and kept in irons and finally thrown off the ship in Hawaii, was the nephew of the Prime Minister. Vancouver had risked the wrath of the most powerful in the land, no matter his reasons for dismissing Pitt. Vancouver, as master of his ship, had absolute authority. Upon his shoulders lay the heaviest of responsibilities, and many a ship's master felt he was ultimately answerable only to God.

The problem for Vancouver was that on returning to the shores of Britain, he descended from this elevated position to being a mere naval captain on leave. The balance of power and influence swung heavily the other way. Ill and failing, Vancouver did not stand a chance. The logs of the expedition lodged at the Admiralty somehow had all references to Pitt's behaviour and sacking in Hawaii removed. Menzies' journal stops abruptly along with many others. Only the most powerful in the land would have been able to erase such history. In the meantime Vancouver went home to King's Lynn to finish writing up his account, and there were no further offers to him to command another ship. He had sacrificed his greatest opportunity. Ill and disappointed, he had less than two years to live.

Whether Vancouver had a considerable flaw in his character, which allowed him to lose his temper and judgement, a flaw exacerbated by the grinding responsibility of such a long voyage, or if indeed he was ill for much of the time, is still a matter of debate. Vancouver has been vilified by many, and supported by others. But the facts are simple enough. Vancouver was thirty-eight years old when he returned with what would later emerge to be a remarkable and accurate survey of the area to which he had been assigned. He was the leader of a heroic journey. But he was physically going downhill. Within a couple of years, Vancouver, at just forty years of age, was dead of an undiagnosed illness.

Menzies, on the other hand, was championed by Banks, who petitioned the third Duke of Portland to allow Menzies' salary to be continued while he organised and wrote up his material. Menzies thus spent three satisfying years writing up his finds and cataloguing his collection. Much he documented, but missing were the accounts of many of his wanderings ashore at Monterey, California, his visit to the Abermarle Island on the Galapagos chain, not to mention the long journey on horseback to Santiago in Chile and the banquet from which so many conjectured stories about the monkey puzzle seeds arose. Eventually, though, in 1799, the navy would pay up no longer, and he was forced to return to sea, gaining a commission as a surgeon, bound for the West Indies. But by this time he had dropped out of correspondence with the other widely acclaimed botanists of the era to the extent that Olaf Swartz wrote in 1802 to Dawson Turner – who in turn was to nurture William Hooker – that he had not heard from Menzies for five years.

Menzies' low profile and uncommunicative years appear to have been uncharacteristic. They might well have been the result of a number of factors. Professionally he was concentrating on cataloguing duplicates of his botanical finds for distribution abroad, writing up the anatomy of the sea otter, detailing his knowledge on mosses, and allowing others access to his collections rather than publishing his finds himself. Personally, there were some goings-on within his family in their homeland of Strathtay and Aberfeldy. Menzies' brother James had become involved in a local protest against conscription and had been charged with treason. James managed to evade trial and flee the country, which must have been a relief to his brother, but had anyone in London made a connection between them it would have seriously affected Menzies' career.

From 1799, for the next three long years, Menzies worked in the West Indies. He was appointed as a naval physician in the area, and served variously on the yacht *Princess Augusta*, the frigate *Tamer* and finally the *Sans Pareil*. Dawson Turner kept track of him from time to time, and remarked that Menzies had had no time for botany in the West Indies. In about 1801 he returned home with his health in tatters due to chronic asthma, exacerbated by the tropical heat and foetid atmosphere which prevailed in the 'cockpit', which was the lower deck where the wounded were treated. He asked to be invalided out of the navy on half pay, backing up his negotiations with the Lords Commissioners of the Admiralty with written statements from himself and his superiors detailing how he had been forced to sleep up on deck in order to ease his laboured breathing. His request was finally honoured. He received no further commissions and set himself up in a London medical practice at 6 Chapel Place, Oxford Street. Although now fifty, he had quite recently married. With a naval pension and a medical practice, he became a benign host for the growing band of plant collectors bound for the west coast of America.

David Douglas, who was to become one of the greatest plant collectors of all, and a young Glaswegian surgeon, John Scouler, who were to sail together to the area around present-day Oregon and British Columbia, contacted him, exchanged letters and visited him together. From his hut thousands of miles away on the Columbia River, by Fort Vancouver, Douglas wrote in his journal that his view was of Menzies Island, named after Archibald Menzies. Several years on, in 1834, in one of the last entries before his death, Douglas described how he climbed the mountain on Hawaii: 'I made the journey to the summit of Mowna [Moana] Roa or Long Mountain which afforded me inexpressible delight. This mountain of 13,517 feet elevation is one of the most interesting in the whole world – I am not sure whether the Venerable Menzies ascended or not. I think he did, and the Natives say he did "for the redfaced man who cut the limb of men off and gathered grass" is still much known here.'

Menzies, meanwhile, had retired from his practice and taken a 21-year lease at a cost of £65 per year from a Robert Cantwell for a house at 2 Ladbroke Terrace, Notting Hill, which was then almost rural. It was ideal for a retirement home, and he and his wife enjoyed entertaining a stream of visitors. Ladbroke Terrace became a central meeting club, a clearing house for many of the great botanical names of the day, and retained a Scottish twang for the remaining two decades of Menzies' life.

Mr Ward, a doctor whose invention of the 'Wardian case' – a glassed-in box for transporting plants around the world – transformed the method of protecting and bringing home tender plants, came to dinner with Menzies and his wife. John Smith, the chief gardener at Kew, came to Sunday lunch with Menzies and chatted about ferns. Sir William Hooker, Professor at the University of Glasgow, whom Menzies had implored to apply for the Chair of Botany at Oxford, visited. George Don junior, who had returned a couple of years before from his Horticultural Society of London expedition to West Africa and the West Indies, was snaffled by Menzies to transport specimens to Hooker in Glasgow. And Thomas Drummond – who had taken over the elder George Don's garden in Forfar briefly – sent the elderly Menzies some samples of mosses, which he knew to be his abiding interest.

On 9 February 1842, Menzies, by now a widower and almost ninety years old, was in failing health. He wrote to his solicitor in a shaky hand to request a meeting. Five days later his final will was complete, naming one of his nephews (he had no children of his own), Charles Menzies of Greenfall, by Crieff, Perthshire, as an executor. He left his herbarium within its mahogany cabinets to the Royal Botanic Garden of Edinburgh, and sums of money to all his surviving relatives, his cook and his housekeeper. His assets were considerable, comprising amongst other things a gold watch, silver plate and 6,220 francs in French government stock.

Forty-seven years earlier, Menzies had sailed back to Britain with George Vancouver, angry, frustrated and apprehensive. Vancouver had had scant time for Menzies, and supporters of Vancouver over the years have painted Menzies as a weasel – a disruptive and impudent character who tried to upstage Vancouver. Whether Menzies had mellowed or not we cannot know, but his later acquaintances paint a very different picture. William Hooker described him as a 'Nestor of Botanists . . . Whom I found a most pleasant and kind hearted old man', a description which echoed the impression he made on his many visitors, all of whom used similar words.

Few botanists have had so many plants named after them. He even has one complete genus bearing his name, *Menziesia*, a group of small shrubs with deep purplish-pink bell heather flowers. Then there are: *Arbutus menziesii*, the majestic white-flowered Madroña tree; *Chimaphila menziesii*; *Nemophila menziesii*, the aptly named Baby Blue-eyes; *Nothofagus menziesii*, silver beech; *Penstemon menziesii*, a garden staple; and the pink-flowered *Spirea menziesii*. But the plant which carries perhaps the most appropriate of all memorials for a man who was a kindly magnet and encouraging influence for eager young plant collectors is *Tolmiea menziesii*, also engagingly named the pick-a-back plant, or youth-on-age, because it produces young plants on its leaves.

Menzies died the day after the signing of his will, on 15 February 1842, aged eighty-eight, and was placed beside his wife in the cemetery of Kensal Green, London. As a 'Nestor' to future botanists, and a fatherly figure to the many whom he entertained in his later years, Menzies might have been conscious of his good fortune; he had after all survived to tell his tales and live to a ripe old age. In contrast, George Don, ten years his junior, never left the sanctuary of Britain, but his harsh life left him dead thirty years before Menzies. He too was a guiding force behind so many future botanists. But for Don's young son George junior, that guiding light was far from evident.

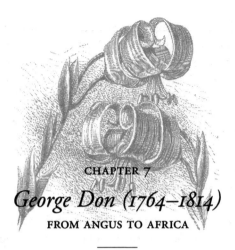

CHAPTER 7

George Don (1764–1814)

FROM ANGUS TO AFRICA

George Don junior, aged sixteen, shuffled miserably behind his father's coffin through the streets of Forfar. Although in the future he would become a successful plant collector, and three of his siblings would achieve success as gardeners, the outlook was bleak for him on that February day in 1814. That the family still had food on the table and a roof over their heads was due largely to the charity of a cousin, General Sir David Leighton of the East India Company Service. A patron of his father's, one James Brodie of Brodie, was among a very few others who supported the family.

Even so, for George's mother, the newly widowed Carolina, it must have been like peering into a bottomless abyss. The market garden on which she and her late husband had struggled to secure the ninety-nine-year lease, and on which they had constructed a family house, still had eighty-two years to run, but was now firmly in the clenched fist of their creditors. The family was virtually destitute. However, if the outlook that grim day appeared desperate, the family situation was about to deteriorate even more sharply.

Sixteen-year-old George, who had been born in 1798, would recall his father's creditors with considerable bile thirty years later. The land on which his inheritance stood was just about to be sold for a fat sum for the building of a new railway station. His father had never recovered from an illness, probably a serious chest infection, which was exacerbated by the climate on the Scottish Highland mountains on which he had wandered gathering plants. His inability to apply himself to the regular business of earning a living hastened things yet further. George senior's death was followed within two weeks by those of both George junior's grandfather Alexander, aged ninety-six, and his only surviving sister.

Outwith the Dons' personal sadness, Britain was breathing a collective sigh of relief that Napoleon's army was in shreds following his disastrous winter campaign in Russia: only 30,000 men had returned from there from an army of 300,000. Napoleon himself was on the brink of abdication. Meanwhile, the erstwhile superintendent of the Apothecaries' Chelsea Physic Garden, Old Meldrum lad William Forsyth, was staving off criticism of his 'plaister', a cure which he had been contracted to produce to save some sickly oak trees, thus securing good wood for new warships. His successor but one at the Garden, William Anderson, was about to take up his appointment. Little could young George have guessed that within a few years he would become foreman gardener there under Anderson's tutelage.

The entire population of the little market town of Forfar turned out to follow the cortege, which appeared to be a great honour in view of the fact that at fifty years old father George had few close friends in the town. Don did not appear to fit in well with a small church-going community. Perhaps his peripatetic life spent wandering the hills and glens in ragged clothes, absorbed by the minutiae of plant life, had little in common with the niceties of parochial, small-town life. His waking hours had been rooted in a deep love of botany, interrupted by sporadic attempts to earn money for the ever-increasing family brood. There is ample evidence of his intelligence and ability to correspond with men educated to a far greater level than he. By nature he was a questioner, and this led to small acts of rebellion. With others in Forfar he set up a local library, equipped with scientific instruments, inspired by the ideals of the radical philosopher and novelist William Godwin.

George junior's grandfather, Alexander Don, was a loyal, supportive member of the family. He appears to have been a gentle, tolerant man who was variously a farm labourer, a currier (preparer of leather), and then a maker of brogues. However, according to his grandson, throughout his life he was 'a great cultivator of flowers for his amusement'. It was he who may have instilled in his sons a love of gardening, plants and the outdoors.

This interest may well have been boosted by the marriage of Alexander's sister Helen to a Mr Miller, gardener at Dupplin Castle, Perthshire. Whether this Miller was related to Philip Miller (1691–1771), legendary gardener at the Chelsea Physic Garden, is unknown, but highly unlikely. But Alexander's brother-in-law Miller not only taught his son, the deceased George, the rudiments of gardening – Miller's son, George's nephew, was also his pupil and indeed succeeded his father at Dupplin. Another of Miller's sons became gardener at Belmont Castle, Forfarshire, home of George Kinloch, a radical laird, who had been so inspired by the ethos of the American Revolution that he had renamed the hamlet within his ownership Washington. Years later its name changed again to Ardler, but it still retains some street names redolent of North America.

As a youth, George Don senior had become a skilled clockmaker, serving an apprenticeship at Dunblane. He then set off for London, working his way down in classic journeyman fashion. He might have eventually found employment as a clockmaker in Fleet Street, but as his son George pointed out, 'his expenditure being much greater than his earnings, my Grandfather [Alexander Don] had to remit him the means of enabling him to return to Forfar where he had by then removed'. 'My father,' George Don junior revealed, 'appears then to have got tired of the Clockmaking business and went to learn Gardening or Horticulture with his cousin, Mr Miller, then gardener at Dupplin Castle, and here he appears to have made his first botanical excursions, and to have formed the acquaintance of the Browns of Perth [market gardeners]'.

This catalogue of jobs and interests, chopping and changing as his interest rose and waned, was to be an accurate and uneasy forecast of George Don's future. Intelligent, impulsive, opinionated, irascible and improvident, he was to be bailed out by his father all his days. While at Dupplin gardening and roaming the countryside botanising, he met his future wife, Carolina, whose middle name, Oliphant, was the same as the owners of Gask House, in which she was employed. Indeed her first name was the same as her contemporary Carolina Oliphant Lady Nairne, who went on to achieve posthumous fame as a lyricist. Her Jacobite sympathies ran deep, and she is the author of three favourite Scots songs, 'Charlie Is My Darling', 'Will Ye No Come Back Again?' and 'Laird o' Cockpen'.

Carolina Don was a stalwart personality, supportive, free-thinking, and of enduring spirit throughout a life in which dreams were sorely tried by grinding poverty and unremitting disaster and death. Amid their roving lifestyle, she was to bear Don fifteen children, of whom only four sons survived into adulthood. Around the year of their marriage, in the early 1790s, George and Carolina moved to Glasgow, where he again took up clockmaking, but the lure of botanising in the hills brought him into contact with similar plant enthusiasts, one Bailie Austin and Dr Stewart of Luss, with whom he corresponded for the rest of his life. But both had something Don singularly lacked, the ability to be financially comfortable and independent.

In 1797 George and Carolina Don grasped at the opportunity to realise their dream. Somehow they managed to scrape together the downpayment for the ninety-nine-year tenancy of an acre of land called the Dovehillock or Doohillock in Forfar. Written into the agreement was the stricture that he was to construct two houses on the site, which looked westwards onto the boggy marsh that, until it was drained, had been the Loch of Forfar. Build at least one house he did, occupied by his growing family. Some of the fifteen children born to them grew up there. In theory, the Dons should have gone from strength to strength from here.

Forfar was thriving. People were industrious, very few bankruptcies occurred, the turnpike road to Perth was just completed, much scratching of heads went on among the city fathers – a provost, two bailies, and twelve common councillors, elected annually – about the education of the young of the parish, something that was highly regarded in Scotland. Two schools already existed with improvements in the pipeline, but locals wished for more: one solely for the study of Latin grammar, another for learned or foreign languages and one solely for the study of English, writing and arithmetic.

Don grew vegetables in order to support his brood. Worryingly, though, his real passion was to build a large pond which he filled with aquatic plants and fish. As this would have entailed digging a huge area by hand, filling it with clay, and then diverting water carefully into it, it was an extravagant hobby. Even worse for the family finances, he could not resist the call of botanising in the hills round about, and his focus on making a living largely disappeared as he roved for months of the year. At least some of the time, Don employed or apprenticed men in his garden, including Donald Munro, who proceeded to become head gardener at the Horticultural Society in London.

Although he profited little in material terms, his wanderings were certainly not in vain. He found many previously unknown species of plant, such as *Eriophorum*, now known as *Scirpus hudsonianus*, a rough grass, and plants which he prophesied would vanish from the wild in a few years, as indeed they did. He also made friends with and won the respect of eminent figures in the botany world.

While in Glasgow in the early years of his marriage, the 1790s, he had met and botanised with John Mackay. Having trained as a gardener at the well-established and well-known Dickson Seed Company in Leith Walk, Edinburgh, Mackay then became head gardener at the Royal Botanic Garden in Edinburgh in 1800. Mackay appears an altogether different character from Don, calmer, more organised and focused. Don all too often collected plants without noting down or remembering where they came from. He did not hesitate to take issue with any botanist he thought was publishing facts in error. He would lose track of time, trudging over moorland for twelve-hour stretches, sleeping in a plaid and eating his bag of oatmeal. Once he turned up at the manse at St Vigeans to be greeted by his friend the minister. When he found out to his surprise that it was the Sabbath he exclaimed, 'Man, I have lost count, but if I had my face and hands washed, I would gang to the Kirk too.' Shown to a bedroom for a wash, he never reappeared, having fallen asleep within minutes. Strong, fit and enthusiastic, Don contrasted with the more cerebral Mackay, who was to die at the age of twenty-seven after a lingering illness. Don was distraught. As his young friend lay abed dying in Edinburgh, Don brought plants and mosses in a vain effort to distract him, and give him hope to live.

He might have reminded Mackay of a trip to Skye when Mackay, searching for Don, located him without trouble. Don was all too easily located, through the simple expedient of questioning local crofters who had found his behaviour so extraordinary and eccentric as to be noteworthy. Don would scramble up rocks wielding a homemade staff topped by an iron hook with which he would tweak down plants. The locals viewed this behaviour with extreme puzzlement. After all, these plants were to them simple, common weeds.

Don accumulated, by default, many good acquaintances. He was never a man to curry favour with friends for their money, influence or titles. But plants such as he found on Skye, Glen Clova and Restenneth Moss gave him ample reason to enter into extensive correspondence with such influential men as James Smith and Dr Goodenough, who was later to become the Bishop of Carlisle.

The bishop and Smith also wrote to each other remarking on Don's zeal, amusing themselves in a slightly patrician way at his endless enquiries and exhortations for prompt replies. These lengthy and frequent letters from Don were more than likely written at the expense of his garden. Although he earned a modest amount from sending specimens to his patrons, his neglected garden was earning him only a pittance. Don was also in constant correspondence with Brodie, whose herbarium went to the Royal Botanic Garden in Edinburgh, although Don appeared to guard his own collection under his own firm thumb at Doohillock, and for most of his lifetime was in regular contact with a number of eminent botanists including Sir William Jackson Hooker who became Professor of Botany at the University of Glasgow and Director of the Royal Botanic Gardens at Kew.

In 1804, James Edward Smith and Brodie pushed Don's name forward to replace the recently deceased Mackay at the Royal Botanic Garden in Edinburgh, and Don duly took up the post immediately. But for the Don family, living on Leith Walk, the Edinburgh years brought mixed fortune. Within a few years Don was to be suspected of claiming Mackay's discoveries for his own. Their youngest son, named James Brodie in the family benefactor's honour, died. And rumours circulated constantly about Don's gardening prowess.

His relationship with Professor Rutherford, the Regius Keeper (Regius being the title given when the post was granted by the sovereign) of the Botanic Garden, became strained. Perhaps Don was a mite too knowledgeable about his subject, and Rutherford felt threatened by this socially inferior know-all. Don was apt to impart botanical statements with dogmatic determination, and his tactlessness earned him little liking. Rutherford might well have felt in part that his position was usurped. Don had little experience of hothouse plants, which was necessary for the post. In the meantime, his long-suffering father was left in charge of the garden at Doohillock in Forfar, from which a meagre income failed to top up the wage earned by Don in Edinburgh. On the plus side, Don was elected an associate of the highly prestigious Linnean Society, among whose illustrious members were the aristocracy, and eminent naturalists, such as Sir Joseph Banks. On the negative side, an ever growing family had to be kept on his pay, which was £40 a year. In comparison, the local church minister's stipend in Forfar fifteen years earlier had been £100 plus a house, garden and glebe (fields) of seven acres for his horses and milking cows. A manservant on a farm received £12, plus house, a small amount of ground for his cow, and vegetables, beer and oatmeal. This might well have been adequate if the Doohillock nursery had shown a profit. Unfortunately, this never appeared to happen. Struggling financially as the family did, Don nevertheless decided to hold on to his post at the Botanic Garden for four years.

Perhaps it was the restrictive institutional life that made Don leave, or the lack of pay, or his inexperience with hothouse plants; whichever, he returned to Forfar, having usefully added medicine to his skills. Although he hadn't actually completed his course, he successfully gathered patients and became partially occupied with medicine. More often than not, though, he was off exploring the countryside for plants. It was generally realised that when a doctor was required in the thriving area of Forfar, Don was usually away. His patients wasted no time in finding an alternative. His nursery and his finances became more and more impoverished. Despite these straits, if there was any spare time, it was devoted to extensive correspondence.

Letters flew out from Doohillock to James Edward Smith, President of the Linnean Society, at his home at Norwich, enclosing specimens of plants and mosses – always detailing where he found them, such as three miles from Forfar. If a reply was not forthcoming, a response was requested in no uncertain terms. On 12 December 1803, no doubt frustrated by the winter conditions, Don complained to Dr Smith: 'I wrote to you several letters at different periods, but have not received any answers at least to any of my former queries, which I am glad to have as soon as possible as I am very anxious about them.' On 18 May 1809, he sent off more plants and again on 20 May. Finally, on 25 June, he sent 'a specimen of a rose, discovered in Forfar some years ago by a Mr Templeton, appears to me to be a different one [from any he had ever seen] – it is from a bush 3 feet high – remarkable petals oblong and truncated and of a beautiful blush colour – I send a young shoot.'

The bishop found the deluge of letters mildly irritating from time to time, and wrote to Smith on 4 July that 'I cannot be but struck with the simplicity of his style and the acceptance of his remarks, and I cry out oh??? George Don! Save!?' A week later he wrote '*Ecce itirum Crispinus*! What an indefatigable creature is this Don!' By August, the flow of letters and specimens drove the bishop to write, 'When I see his accounts and observations I feel I know little and really think I know nothing. Where have all these things been since the days of Adam that no-one should have noticed them before.' But he had to acknowledge that Don's endless correspondence was fruitful. Don succeeded in finding plants. On 21 July the bishop wrote again: 'What an indefatigable fellow is G. Don. There is no end to his researches. The world almost expected that he had saved seeds of foreigners [i.e. had procured seeds from foreign parts, but pretended to have found them in Scotland]. But I would willingly suppose he is above such artifices.' By 28 July 1810, he wrote again: 'Dear Sir, Wonderful Don! What things he finds! And I send you two more packets from this wonderful man [underlined].' Two days later, at the end of July, the bishop refers to the 'Indefatigable Don, he wants my opinion of his new *Avens pubescens*, but really I am fearful of hazarding a sentence upon this nice plant and willingly submit to other heads, – yours and his.' Don was still firing off letters in April 1813, dispatching specimens from the summit of 'Ben McDawie [Ben Macdhui] the high mountain at the head of the River Dee in Aberdeenshire, which I have never seen elsewhere.' By the summer of that year he wrote to tell the bishop that 'I have just received the melancholy intelligence of the death of my worthy friend, Mr Dawn of Cambridge – he is the last of my personal acquaintance which I acquired during my stay in London . . . I was thirty years acquainted . . . and have lost a most worthy friend and faithful correspondent.' But Don lost no time in getting back on track, which was to keep up the correspondence about plants. He quotes 'Let us labour while it is today / For the night cometh when no man can work', then continues with a list of plants for perusal. It could have been his own epitaph; he died on 15 January 1814 after six weeks of illness.

Throughout his life, Don's methods of collecting were the subject of constant questioning and bickering. As already mentioned, he was suspected of claiming Mackay's discoveries for his own, and it was also suggested that he extracted all the plants from the vicinity of Auchmithie so that his successor at Doohillock, Thomas Drummond, could not find a single specimen. Some thought he had brought plants back to Doohillock from the wild, had lost track of where he had found them and so just passed them off as his own breeding. Others were kinder in their judgement: he had not noted down exactly from where he had gathered specimens, but this was not surprising as they were mostly from largely unexplored and little-mapped areas of the Highlands. And so the arguments raged. Don was irascible, didn't kowtow to those wealthier or higher born, was well known by many correspondents who had been at the receiving end of his lashing tongue, and pursued his plant collecting with almost evangelical fervour while his health and finances ebbed.

His improvident and independent nature had produced a double-edged result. He was bankrupt but he had left behind a vast legacy. He had unwittingly demonstrated that a poor man could still become a forceful collector and a correspondent with the most eminent men of his day. He had, unknown to him, inspired a generation of collectors from all backgrounds. Within eight years of Don's death, his son George was the collector chosen by the Horticultural Society of London to go to West Africa. David Douglas, Archibald Menzies and John Jeffrey are among the many who referred to Don fulsomely. There is not a memorial to him in stone, but his spirit was carried within humble Scotsmen who were to feel that they too could take on the world. As Don was buried, and young George railed against the world, Thomas Drummond slipped into his garden and used the Forfar soil at Doohillock as a springboard for ventures in the west; not the west of Scotland, Don's beloved hunting grounds, but the more rugged and challenging west coast of North America.

CHAPTER 8

George Don Junior (1798–1856)

SUGAR, SLAVERY AND SEEDS

———

George Don senior may have left the family destitute, but his reputation attracted eminent men to his graveside, one of whom, Dr Patrick Neill, took it upon himself to keep an eye on the surviving sons. Although qualified as a lawyer, Neill was a man of many parts and having usefully inherited his father's printing works, founded in 1749, was able to indulge his interests. He was secretary of the Caledonian Horticultural Society for forty-one years and had originally met Don one memorable day in 1797. 'On reaching Forfar towards evening, I soon found Don's garden, and entering, inquired of a very rough-looking person with a spade in his hand, whom I took for a workman, whether Mr Don was at home. The answer was "Why, Sir, I am all that you will get for him."' Their friendship was sealed that evening in the local inn, after an excursion to an area of moss where Don pointed out the very rare *Eriophorum alpinum*, predicting that it would be extinct in a few years, as the bog-land was being drained. How correct he was.

However, Neill now found George junior and his younger brother David to be 'most opposite characters'. George was determined to carry on his father's work, despite Neill's reservations, but years later, when the two were on much better terms, George was able to acknowledge freely to Neill that he had been too young and inexperienced for the task. The passage of years, however, had not diminished the bitterness he felt towards the way his father had been treated by the burghers of Forfar.

Within a year of their father's death, both the brothers, George and David, had left home and were working at the great training ground of Dickson's Nursery in Edinburgh, while their mother had also given up the fight to retain Doohillock, and had gone off to live in Newburgh in Fife, the place where her late husband had made his last botanical foray and where he had been taken terminally ill. With her she took her other three sons, Patrick, James and Charles.

George senior had certainly possessed determination, if lacking in self-discipline. His son George, however, was more focused and his pugnacious determination was already proving successful. After only a short time with Dickson's he moved to London and, by the early 1820s, was foreman at the Chelsea Physic Garden under William Anderson.

Like all the senior employees there, Anderson favoured the Scottish lads. Young George's early education at the proud schools of Forfar would have impressed Anderson, as well as his love of plants, perhaps his father's influence, a strong working knowledge of gardening and botany, and a burning desire to assuage the wrongs to his father's memory.

William Anderson was a rough diamond with a soft heart: a 'man of rather rough manners . . . of a generous disposition, [who] did many kind acts for necessitous friends . . . he was tall and burly with ordinary coarse and old fashioned style of dress.' Having started out as a labourer at the Chelsea Physic Garden, he was an energetic man who rose rapidly through the ranks. A few years after his appointment, and around the time of Don's arrival, he was demanding a new pit for alpine plants and, then in 1822, a cistern for the cultivation of water lilies, as well as cherishing tricky arrivals in the *Geranium* and *Pelargonium* family. *Pelargonium cordifolium* had been introduced from South Africa by Francis Masson in 1774 and contemporary nurseryman Robert Sweet noted that Anderson cultivated 'many of the old species still remaining that are nearly lost to modern collections'.

Anderson was also a survivor of the sharp-tongued John Lindley, appointed to the Chelsea Physic Garden as *Praefectus Horti*, more senior to Anderson, and indeed in charge of the garden overall. Lindley was ever manoeuvring to rid himself of Anderson, criticising him endlessly. But the wily Anderson was still in post throughout this onslaught, which lasted from Lindley's appointment in 1836 until Anderson's death in 1846.

But Anderson possessed a forgiving nature. Despite the difficulties of his tenure, he left a legacy to the Chelsea Physic Garden on his death, perhaps only made possible by his lifelong bachelorhood. His £100 per annum wage was, in later years, topped up with a bonus of £20 for 'his great attention and care' in his work.

Very soon after arriving at Chelsea, Don was singled out by the Society as a likely candidate for plant hunting and was engaged by the Horticultural Society to sail on the *Iphigenia*, in November 1821, to the coast of Sierra Leone. His wage would be £1 per week for the first few months, from August to November, but thereafter would rise to £2 per week. For this he was expected to risk his health and very possibly his life. The Horticultural Society directed him to 'hunt for plants in three continents and scour the islands in between' and make sure the plants were brought back alive. He was required to carry out this demand by dealing with natives who spoke many different languages, and, generally, by working hard from November 1821 until, God willing, his safe return in February 1823.

The trip had been arranged by Captain Edward Sabine of the Royal Artillery, brother of the Secretary of the Horticultural Society. Commodore Sir Robert Mends was to command the ship.

Although Don never mentioned the orders for the ship, he must have been well aware of the nature of its destinations, and the reason for His Majesty's ship to be calling at such countries and islands on the west coast of Africa. As a true son of his father, he might also have felt a certain quiet satisfaction in his position as a spectator on history. Not lost on him might also have been the irony of searching out 'tropical plants' for the hothouses of the well-to-do in Great Britain, the wealth of whom might have been created by the very sugar gracing their tables. To enjoy this wealth was directly or indirectly dependent on slavery. Here he was, on board a ship dedicated to eliminating the slave trade, and yet, at the same time, the keeping of slaves in the West Indies by British landowners was showing few signs of being eliminated. Wilberforce might be raging in Westminster, but wealth within Britain still depended on the inhuman exploitation of black Africans.

As Don studied the list of instructions from the Society for what was expected of him, he must have wondered if he indeed would fulfil his father's ambitions and become a botanist of note. The list ran to five and a half pages, written in a slightly haphazard way, as though additional instructions had been tacked on as they occurred to the Society's committee. He was to be under the control of Captain Sabine at all times. He was to return punctually to the ship at all times, and in the unlikely event of requiring further supplies either for himself personally or in the course of his work, he was to apply to Sabine or the captain of the *Iphigenia*. He was to keep an eagle-eye open for the types of plants that were not well known in 'England' (how that definition must have grated in a Scottish lad), and make quite sure that if they were growing in gardens of the places he was visiting, he was to note carefully how the local gardener was looking after them, and for what use the plant was desired. 'Such as is likely to [be] useful as fruits or esculents are of the first importance and next to those plants which will be esteemed in our gardens, for their beauty or singularity . . . however, amongst the great mass of plants besides it is desirable that especial attention is paid to palms, to bulbous plants and to the orchidaceous tribes.' No matter that the plants 'will be of course native to hot climates, but the Society [is] especially desirous to obtain these tender plants'.

So much for the plants he was to search for. He was also charged with keeping a fair journal of his 'proceedings and observations', gathering and drying specimens, labelling everything very thoroughly, writing to the Society from the first British port on his return, seeing that his specimens reached the Society on his return with alacrity and, finally, noting well that everything in his possession, every plant and seed, every skinned bird and his journals were the property of the Society. Very little was left to chance.

'Collecting' extended to people, who were also featured on the 'to do' list, popped in as point eleven on his list of instructions. Although he was not to be given letters of introduction for the Society's correspondents around the globe to whom he was to make himself known, his letter of instruction would be deemed sufficient to open up the doors. A British society was confident of its important place in the world, and its very name, it was sure, would provide an open-sesame for its chosen representatives as well as a source of attraction for new contacts. This was an ambitious social remit for a lad from Forfar whose gardening know-how had been fine-tuned by the energetic William Anderson, but whose social skills, acquisition of new languages and seafaring legs were to be learned only as he went along.

Whether he had ever been to sea before is a moot point. Travel from Edinburgh to London was usually on board the small merchant ships plying from Leith, close to Edinburgh (and now part of the city), to various ports of call on the English coast, and finally reaching the Thames for disembarkation. The Society equipped their new recruit with an 'outfit' costing £39 19s 8d, within four pence of his entire year's wage for the trip. The outfit also included knives and scissors (£1 8s 2d), camphorated spirits of wine (17s 4d), coach hire with boxes to Charing Cross (2s 0d), a massive bill for stationery (£12 14s 7d), and a trunk from Parke and Co. (£1 7s 0d).

Into the trunk went such items as one hundred pens, a box of wafers, two quires of foolscap paper, two pieces of India rubber and a small letter case deemed essential to protect and keep his instructions and passport watertight. Further items included a dozen pairs of pliers to cut specimens, one dozen each of ink powders and pencils, two boxes containing bottles for fruit and one and a half gallons of spirits to preserve them, eight pruning knives, two triple-magnifiers, two boxes with partitions for insects, a pair of forceps for catching insects, preserving powder for birds and various instruments for skinning birds.

Material for these tasks included Pearson's *Synopsis Planetarium*, two volumes of *Hortus Suburbanus* by nurseryman Robert Sweet, who was also in the midst of penning five volumes on the *Geraniaceae*, which he managed to complete between 1820 and 1828, and finally, a tome on taxidermy by Bullock. All of this was quite an ambitious homework project to be carried out on the high seas. Dressed in his new outfit, and with trunk packed with homework and tools of the collecting trade, the young George Don was sent off into the world, charged with bringing back plants for the profits of the Society and establishing new contacts from West Africa to North America and points in between.

Don was to sail on HMS *Iphigenia*. Such frigates were equipped with from thirty-two to forty guns mounted on a single gun deck. The *Iphigenia* possessed thirty-six.

For some time Great Britain had been at war with a succession of enemies, some of whom later became allies. Many vessels during the war spent their sea-time in cruising or escorting convoys of merchant ships across miles of empty ocean. In this humdrum existence their main enemy was the violence of the sea. Rarely, if ever, would they encounter an enemy warship or privateer.

After the fall of Napoleon in 1815, most of the battle fleets were laid up 'in ordinary' for the remainder of their days, but large numbers were also re-employed for anti-slavery patrol around the African coast where small, fast vessels were essential in order to catch slavers. The *Iphigenia* was such a ship, and this was her last active trip, her keel having been laid down many years before, in 1808, at Chatham. She had already traversed the Atlantic several

times, experiencing some exciting days. Some were not entirely
without farce. In August 1809, while cruising off Mauritius, she
had accidentally run into the *Boadicea*, losing her bowsprit and fore-
mast. The following night she ran aground under a battery where
she was exposed to heavy fire. Not thinking that she could be saved,
the senior officer of the squadron ordered her to be burnt; but by
throwing some guns overboard, the crew eventually succeeded in
getting her off. However, because of the damage she had to dock
in Bombay for repairs.

In previous trips she had carried 255 men and boys, and at least
a couple of eminent naval commanders on board. In 1818, Captain
Hyde Parker had taken her from Quebec to Jamaica and the
Mediterranean, suffering the loss of eighty-five officers and men
to fever during her time in the West Indies.

Fever was capable of claiming far more casualties than active
warfare. Yellow fever, or 'Yellow Jack' as it was called, would
frequently kill three-quarters of a ship's company and leave the
survivors too weak to work the sails. During 1816, the *Childers,* in
the period of a month in the West Indies, lost several officers and
thirty-five men from fever in addition to five pursers appointed
in succession, and, in 1820, the 26-gun *Tamar* arrived in Halifax
from Jamaica with scarcely enough men to bring her into harbour,
her captain Arthur Snow and seventy-five of her crew having died
during the voyage. It was not until the beginning of the twentieth
century that it was discovered that yellow fever was carried by
a mosquito.

Don's big adventure had started off so well. He left London by
stagecoach for Portsmouth on 13 November 1821, paying out £1 10s,
plus another 4s to the coachman and guards, not to mention 2s
for refreshments on the road and, as an afterthought, a purchase
of three pairs of coarse gloves and a tape for hanging knives, all
for a cost of 6s. He checked in at the George Hotel, pleased to
find myrtles and some laurels 'the very best he had ever seen'. By
17 November, he was on board the HMS *Iphigenia*, and finally sailed
on 20 November at two o'clock in the afternoon. A mere three days

later, 'exhausted by sea sickness', he went ashore to look round the gardens, but was back on board and refreshed within a day.

By 6 December, he had proceeded on foot as far as Tor Bay (adjacent to present day Paignton), then the following day to Brixham to study gardens, while his mentor Sabine was still at Teignmouth, further up the coast. On the 24th, he and some officers, accompanied by two prisoners, decided to take the last boat out at three o'clock in the afternoon, despite the fact that all the other officers for the ship were back on board by now. However, such a storm blew up that they had to remain ashore, while the ship sailed on down the coast.

Then followed a chase to catch the ship. On 27 December, they left Brixham in an effort to catch the *Iphigenia* at Plymouth, but heard the ship was now at Tor Bay. Two days later they left around eight o'clock in the morning on a small pygmy boat to sail to Tor Bay, but the seas were too strong and the commander from whom they had been given the boat ordered them back. Don and a midshipman decided to walk the 32 miles to catch the ship, but 23 miles on at Totness, almost within sight of Tor Bay, they gave up for the night, soaked to the skin. Realising that the westerly wind would prevent the ship from sailing, they resumed their journey the next day by 'car'. On 30 December, they were safely back on board. On the next day, as the *Iphigenia* proceeded down the coast, the original pygmy boat was spotted making for the ship, carrying the officers and prisoners. Two hours later, with all on board, they set off, but only as far as Plymouth, where they anchored again and saluted the Admiral before finally setting off for Madeira. Don was to bill the Society for 'things being left behind cost £5 6 shillings and 3 pence'.

Don was clearly upset and rattled by this mix-up. This was a world of which he had almost no knowledge. Feeling the effects of seasickness on his first major sea voyage at the worst time of the year, he must have been dreading the stormy Bay of Biscay in January. Already he had walked miles in pouring rain, and was anxious about his failure to live up to expectations even before he had left the shores of Britain. Luckily, neither Sir Robert Mends nor Sabine made much of this particular incident. But unknown to Don himself, Sabine was about to take issue with Mends, on Don's behalf.

It was all to do with Don's accommodation, which was a hammock slung under the half-deck. Don had to eat in the gunners' cabin, and the trunk he had so carefully packed with his brand new tools and paper was stowed away on the lower deck. But the issue had less to do with Sabine's concern about Don's comfort or about any insult to Don's status, but, rather, much more to do with ensuring conditions that would allow the maximum output from his protégé. It transpired that Don had no place to read or write, nor enough space to sort out his specimens. In short, his accommodation was totally inadequate to fulfil the Society's high expectations. Sabine was hopeful that better quarters would be forthcoming after their arrival at Madeira, where Don was to be landing and would be officially on duty, but he was thoroughly exercised about just how to go about this delicate issue. The problem was not one of naval overcrowding, or military hardware absorbing free space. In Sabine's view, the problem was Commodore Mends' family who were on board for this last voyage of the *Iphigenia*. Their presence was taking up the valuable space.

Sabine carefully crafted a letter, even to the extent of issuing a veiled rebuke, to elicit an immediate response from Mends. This was on a small point of principle, of accommodating a lowly gardener abruptly elevated, though not very far in social terms, to botanic explorer. Indeed Don was certainly a long way from being treated as a first-class passenger. But it was far more than this, and Mends knew it. Sabine hinted that if a cabin was not found, Don might as well be sent home from Sierra Leone.

This might have rankled with Mends who, like most naval men of the time, knew that patronage and influence were vital in advancement. He must have quickly reasoned that the Horticultural Society members, men of substance and highly influential in the heart of the nation's capital and with easy access to the decision-makers, might be dismayed, and his future career might be impaired.

Mends responded that he had not realised that he had to offer more than 'a package' on the ship. A new cabin was promptly located for Don, as well as 'a man to accompany him on all his excursions'. Furthermore, Mends was anxious to let it be known that he wanted to assist Don, and to promote 'his [Mends'] goodwill to the Society'. Honour was satisfied all round, and Sabine withdrew his letter, thereby preventing it from ever reaching the Society's eye.

In all this, and despite Don's dash around the south-west of England when he chased the ever-moving ship, Sabine was able to reassure the Society's committee that Don was not only behaving very well under such circumstances – in his hammock space – but also that he should have 'at least the same rank as a midshipman'.

Oblivious to such wrangling, and no doubt also juggling with the effects of seasickness and diagrams of taxidermy, Don sounded like a child finally arriving at the beach on holiday when he reported breathlessly on the wonders of Funchal, Madeira, on 11 January 1822. In the first of his four days there, he recorded that 'we went ashore, after taking our things through the Customs House, and stayed ashore . . . The day being very wet and likewise being the middle of their winter I did not see half what I should have seen . . . [observed] *Nerium oleander* (luxuriously pink flowering shrub) in flower, along with capsicums, various, in beds interspersed with English vegetables. Here are also hedges of *Fuchsia coccinea* intermixed with *Aloe arborescens* and *Geranium.*'

He wondered at the narrow streets, 'very confined and which in my opinion render it very unwholesome'. The following day he viewed Mr Blackburn's garden at Palmara, around a mile from Funchal, the 'best cultivated garden of the islands – finest groves of oranges, lemons, guavas, plantations of coffee and I believe the finest vineyard', which was in accordance as he was 'an opulent merchant and a passenger on the *Iphigenia*'.

Perhaps due to his newly elevated status on the ship, and Sabine's determination not only to ensure that he was treated as an important passenger on board but also to get him out working, Don found himself accompanying Sabine, Mends, Mr Blackburn and the ship's surgeon, Whitelaw, on horseback to one of the summits of the islands. It was a surprise to find snow a foot deep and 'very dangerous on descent'. No doubt he felt justified in billing the Society for 'the hire of a horse to accompany Captain Sabine to the peak, £1 6s, and washing sheets and towels 4s 6d', the first of many such laundry bills.

Among Madeira's native plants were *Tamarindus indica* (Tamarind tree), 'the fruit of which never comes to perfection'; *Arachis hypogaea* (peanuts) planted in beds in the garden and *Dioscorea* (yams), a native of the island 'which Mr Blackburn thinks equal to any of the West Indies ones'.

With that, they sailed to Tenerife. It was a whistle-stop tour. By three o'clock in the afternoon when the ship's gun fired the shots to summon everybody to come back on board, Don had collected a handful of a type of cornflower, a *Centaurea*, some lavender (very common) and an *Olearia* (daisy bush) about thirty feet high, which he reckoned was the same as Mr Anderson had received from Mexico.

Shipboard life was frustrating for a bright, young botanist who wanted to prove himself, especially when the ship weighed anchor and remained becalmed within sight of the islands for four days.

As they proceeded south between the islands of Cape Verde, St Antonia and St Vincent, there was a flurry of excitement as an American ship was spotted which could have been a slave vessel. Immediately a boat was launched to intercept it but the ship was found to be only a whaler. But mainly the days passed uneventfully. 'Today, and for first time saw flying fish', he reported tersely. Thankfully they landed at St Jago, and he went ashore with Captain Sabine to the valley of Le Trinidad. Although the parched islands, with virtually not a green leaf to be seen, might have justifiably dimmed his excitement at being ashore, he was able to delight in seeing 'the finest cotton I ever did see'. This was perhaps a strange remark for a Scottish lad, as he cannot have viewed much cotton. But before going back on board at about four o'clock in the afternoon, when he observed the temperature to be 72° Fahrenheit, he had made a list of plants.

The next morning he went on shore with a fishing party whose success with casting nets twice and hauling in substantial quantities of fish was hardly matched by his plant collecting. Only *Convolvulus* was to be seen. On the way he had also observed 'two of the most miserable huts, thatched with dates palm leaves'. He promptly shot a few birds for his collection and pocketed a few insects.

The following day the coast of Africa came in sight which he thought appeared to be very 'low'.

By the time the ship dropped anchor off the River Gozee (present-day Gambia), Don must have begun to question just how much freedom he was to be allowed to venture ashore. Either there was not enough room for him on the rowing boats which the officers used, or the ship had to anchor too far offshore to allow a rowing boat time to reach the shore and get back in a day. With justification, Don's diaries hinted at his exasperation and concern that he was not fulfilling his requirement for the Horticultural Society. He noted that Captain Sabine had returned from an island at the mouth of the river with specimens of *Argemone mexicana* (a yellow-flowered annual herb, the sap of which is used in medicine), which he declared to be the only vegetation to be seen.

When Don finally got ashore, he found the ground just as Sabine had described it, so parched that finding any specimens to bring back was difficult. 'I saw only *Convolvulus*, *Musa* [bananas], *Hibiscus*, *Epidendrum* [orchids], but I understand several English country plants are cultivated here such as radishes and others that come to perfection.'

By 6 February, they had anchored off Cape Roro. Don's anxiety to get ashore and pursue his duties in plant collecting was tempered the following day when the first boat to be sent off was overpowered by 'natives with bows and arrows and obliged to retreat'. Ten days later, with Don aboard, the same boats ran aground, were then refloated, and rapidly dispatched to check out a slave ship lying under the Portuguese fort. Returning to report that it was indeed a slave ship, the small boats were sent off to capture the vessel.

Arriving in Sierra Leone a few days later, Don was determined that his best chance of plant collecting was to transfer his belongings ashore and take some lodgings for a few days. He must have been advised to make contact with the governor and seek a room there but, as the governor was away, he sought lodgings in the town. A 'most immodest sum' was required, however, and his expenses were too meagre for even a couple of nights. His frustration at being at the beck and call of the navy and unable to spend time plant hunting was beginning to concern him, but he passed the time usefully by studying the Horticultural Society's instructions.

List of plants wanted from Sierra Leone:

Epedendri [orchids], Air plants. There is a large tree on the road to Congotown left to hand, after you pass the bridge, where air plants are growing in great profusion. There are a few sorts there which have not been able to have been brought to England; and there is also on the left on the road to Mr Macaulay's farm, a tree of the country's cherry with two strange sorts [of cherries] upon it. They have cylindrical leaves. They are also to be got in many other places by cutting down the tree.

Large tree.

Having the appearance of Laburnum grows plentifully in the woods, flowering about the end of April. Seeds of this.

Monkey apple.

Young plants of this can be imported with more success than seeds. Seeds may be tried in earth.

The clematis near Governor's town.

Such instructions, some of which were surprisingly accurate but most exceedingly vague, must have justifiably made Don anxious. The Society's scribes back in London must have imagined

Don skipping ashore and following instructions rather like an orienteering competition.

He read on.

Osbeckia. [Beautiful shrubs with deep pink or purple flowers]

Beautiful in the low lands near Freetown. Flowers in great profusion about the beginning of April. The flowers are blue. The roots of this plant are used in the making of Loosoo beer. It is known to Dr Barry. Seeds may be sown in earth but roots are more desirable.

Also *Ixia*, bulbs highly acceptable. [Lily-like bulbous flowers, now commonly available]

Freetown had been a British colony since 1808. The British Navy were using it as a base for their patrols along Africa's coastline, and freed slaves from British territory in the Caribbean were settled there. The first 400 of them had arrived in 1787, and they had been expected to support themselves by farming, but many had turned to trade. Since 1815, Freetown was a centre for slaves rescued from slave ships by the British Navy. Hundreds of ships hauling slaves were seized, while some got through, taking slaves to be sold to sugar planters in Cuba and Brazil. British merchants were pursuing legitimate trade along Africa's Atlantic coast. From Senegal, the British were acquiring a hardened resin of the *Acacia* tree, used for dyes in its textile factories. They were acquiring groundnuts from Guinea, gold from Asante and palm oil, which was harvested on African-owned plantations using slave labour. Palm oil was Africa's foremost export to the Europeans, the oil used for lubricating their machines.

Finally, matters improved for Don. Armed with all his books and collecting apparatus, he went ashore with Sabine and was given a mule to convey himself to the governor's farm, some distance from the official residence, where he was based in an adjoining farmhouse. It took him two hours winding his way up the road lined with fine apple trees to reach the governor's one-acre garden which was full of European vegetables. He endured a couple of noisy nights, being kept awake by howling 'wild animals and mosquitoes'. He hired two local men to travel with him, to carry his bags and boxes. He went a little distance from Freetown to 'White Man's Bay' where he saw a curious, delicate mimosa, and so many pineapples that he had difficulty stepping over them.

Like gardeners all over the world, 'that great Masonic brotherhood' described by Vita Sackville-West, he encountered unlikely tillers of soil. 'Up the mountain, I found a small but very neat farm, occupied by one of the first Nova Leatean settlers (an old woman and the only one I believe now remaining of the whole colony). She cultivates cassava, sweet potatoes and great quantities of pineapples from which she makes a very pleasant type of wine. She has fields of *Maranta arundinacea* (arrowroot) in her small garden as well as rosemary, thyme and fennel, and the whole area is covered with pineapples.' Don had stumbled across one of the strangest relics of immigration.

In 1792, a determined Afro-American called Thomas Peters, who had been born into slavery in North Carolina, arrived in Sierra Leone with a party of freed black slaves to found a new life. During the American Revolution, Peters had run away and joined the British Army, becoming a sergeant in the Black Pioneers. Of course he was on the wrong side and he and hundreds of other freed slaves spent seven years in Canada, where the land that they had been promised never materialised. Cold, and eking out an existence in Nova Scotia, Peters had been waiting to be offered a better deal from the British. Although poor and with limited education he somehow had managed to get himself over the Atlantic and to England. It was an enormously courageous move, and if he had been captured he could well have ended up back in the United States in slavery. However, once in London he convinced the Sierra Leone Company to dispatch ships to bring the rest of his fellow refugees in Nova Scotia to Sierra Leone. Returning to Nova Scotia he managed to persuade more than 1,100 former slaves to join him. He then founded Freetown in 1792 with his Nova Scotians. It should have been an idyllic settlement, but very quickly everything went wrong. The white commander opposed their wish to self-rule, insisting on taking over himself. Peters contracted malaria in the first rainy season and died the same year as they had arrived.

If Don had fretted a couple of weeks about his confinement on the ship, now ashore he found that from the end of February until the beginning of April his feet hardly hit the ground. He chopped down trees to secure orchids, was eye-level with the baboons and chimpanzees of the mountains, and strayed as far as Regent's Town. He searched, as instructed, for seeds of the 'red weather' tree, but with no success, but did find strange 'rough skinned Plums', a triumph not without danger as the area was infested with snakes, 'a green and yellow one being exceptionally dangerous'. He found a 'butter and tallow' tree (probably *Pentadesma butyracea*, or commonly called the shea nut, from which a butter-like substance is obtained) in flower, and, surprisingly, a type of black plum, which grew in one of the streets on trees that were like chestnuts.

Most usefully of all, Don managed to make excellent contacts with the local European doctors, one of whom, Dr Nichol, he found to be most knowledgeable concerning succulent and medicinal plants. He also accompanied a Dr Barry on hospital visits to Glosher Town, up Lister Hill, a few miles away. There they found a 'captured Negro in miserable conditions, who had been rescued from the slave ship a few days before'. Criss-crossing the area, he found help from a local merchant, more than likely to be Kenneth Macaulay, who was one of the most successful businessmen in the area, and a partner in the firm of Macaulay and Babington, whose vast timber crops were transported back to Europe to provide a ready supply of the hardwoods required by the furniture-makers for the fashionable.

Although still unsuccessful in obtaining seeds of the monkey breadfruit tree, which was one of his prime instructions from the Horticultural Society, luck was frequently on Don's side. On a trip to Wellington town, three miles beyond Freetown, and after a chat over a garden fence, he was given not only refreshments, but a large bowl of *Caladium*, about 12 inches across. (*Caladium* are today popular houseplants with exaggerated heart-shaped, often variegated, leaves.) While dodging tornadoes, and nursing a sore foot, he also stumbled across a fifth type of the species *Combretum*, each of which, he noted, was more beautiful than the last. (*Combretum* carry orange or scarlet flowers shaped like bottle brushes.)

More group outings followed. With the assistance of a local minister from Regent's Town, the Rev. William Johnston, he ascended Sugar Loaf Mountain with a local guide, remarking on the beauty, and the following day he took a boat 30 miles to York settlement to stay with Mr Johnston, who was the superintendent. There he found strange bulbs, growing individually rather than in clumps, and unlike any seen in England, plus 'they never appeared to be further than 100 yards from the sea and the beach'. Once back to base, he had a hectic rush to pack up five boxes containing living plants and seeds which would be sent back to England. Although he eventually managed to identify many of the succulent plants he had found, he was unable to establish the identity of the 'miraculous small berry of the Cape Coast which is eaten with everything, and you can taste it hours afterwards'.

As the ship left Sierra Leone behind, Don was adding up his expenses: '22nd February, 1823, at Sierra Leone. Paid man on board the ship for drying paper 10 shillings. Refreshments from the 22nd to the 25th of February, five shillings.'

A few days later, the ship sailed into the St Thomas Islands, one of the biggest producers of sugar in the world at that time. They were immediately surrounded by canoes laden with *Casado* (a type of pepper from Guinea), along with ducks and hens to be exchanged for tobacco.

They anchored off the town of St Ann de Chaves, where they found wooden houses lining regular streets and had time to breakfast with Mr Fernandez, the commodore appointed as British Vice-consul. Although Sabine refused to allow Don to travel further into the island, he nevertheless found guavas and sweet and sour apples before a tornado drove them back out to sea and fever claimed its next victim, the ship's master-at-arms.

Eventually Sabine relented and, despite the danger of illness, rented 'Fish House' in the middle of the town. Don was now gaining confidence and getting into his stride, and his diary even becomes eloquent. He found a beautiful tree about 100 feet high bearing fruit about 'twice the size of a man's head, the seeds of which the inhabitants boil and eat' he noted, and orange and lime trees within the surrounding forest, which 'are now almost in a wild state'. On the same day he found another beautiful tree, *Pandanus candelabrum*, over 120 feet high, with 'four branches in a whorl divided at the point with several pendulous fruits upon it, but not perfectly ripe'. Triumphant as he was, he and his collection were soaked by the waves on the way back to the boat.

The following day, Don brought his own cot from the ship, so that he could stay at Fish House. He observed that everything possessed something of interest, from great herds of monkeys to trees bearing fruit about the size of a plum.

He travelled three miles inland with two marines and one local man, an exhausting trek because of striding through the long grass. They quenched their thirst by drinking coconut juice, and later walked ten miles along the beach to the local market, marvelling at the beautiful species *Ipomoea*, or morning glory. To obtain much-desired orchids he chopped down the tree which carried the parasitic orchid *in situ*, but often he found the wood was so hard he could not pierce it more than inches so was obliged to abandon the attempt. The next day he went to the top of Convent Hill, surrounded by species of *Ficus* (figs), and the following morning was catching crayfish which Captain Sabine believed to be a new species, and in the afternoon found a laurel resembling cinnamon but with no scent, but hosting 'an immense number of parrakeets'.

He was confined to bed for a day, but the next morning he was up again and set out with two marines for some beautiful plantations where he found bananas, plantains, great quantities of cassava, immense trees of bamboo, limes, oranges and a pomegranate, the slender form of which he supposed was due to the heat. In his search for cinnamon, he wisely took a guide to show him where it was found.

'We set off in an easterly direction, walked at the rate of 4 miles per hour, reached the place after five hours, passed over several mountain ranges, saw several trees of cinnamon, but these', he wrote with masterly understatement, 'were probably imported from Brazil.' Two days later, having also found a curious *Bannisteria* (vigourous type of vine), he set off with a marine and three natives for the island's peak. Before long, the natives disappeared, but Don and the marine proceeded on, breakfasted at 3 p.m. at Ville de Guadeloupe, a small village of about twenty homes, found innumerable interesting species, but not the longed-for orchids 'which would have been so popular in England'.

Back on board the ship, he packed up all his seeds which he had collected in the course of eight days on the island of St Thomas, and just aboard, on 14 June, collapsed with a violent fever. Although ill for nine days and unable even to walk for another week, he was fortunate. A brief entry in his journal on 25 June notes that he is getting stronger every day, but the two marines who had been on shore with him had both died. Don was the sole survivor.

That same week, Captain Sabine's brother, at work in London, was penning a letter from the comfort of his polished mahogany desk at the Horticultural Society in London.

'You seem to have made good use of your time. I never had any doubt. I have received the boxes. You have a fine field before you in Trinidad and the West Indies.'

Sabine's brother went on to explain that owing to 'the public duty of the government to require collectors to form collections of natural history', he reminded Don that, as an important part of his duty, 'it may be desirable to send you on other expeditions where your full duty of collecting will be on yourself'. So Don was clearly to apply himself to taxidermy, even at the expense of plant collecting. It was not so much a rap over the knuckles, as a reminder of just who was in charge.

But he went on to tell Don of 'news of your brother David – Mr Robert Brown has given up the office of librarian and clerk at the Linnean Society, and it was unanimously decided at the council to appoint David in the occupation of which you will find him on your return. Mr Lockhart, Sir Ralph Woodford's gardener, is made a corresponding member of the Society and will attend to your instructions. You are to look out for *Arachis* (peanut) root, then *Caracas* called *apios* – it is sold for food in the markets.'

Arriving at Ascension Island, one of the officers not only directed him to the gardens but also found him transport. It might have been a lowly mule, but he soon found that the incline was precipitous. 'Went ashore at 9.30, started up mountain, and distance of seven to eight miles – road good and level for the first 4 miles then almost perpendicular. Saw only *Portulaca oleracea* [type of succulent] and a species of *Euphorbia* [spurge], the locals eat the former as spinach. There is a beautiful *Hibiscus* which the local people call St Helena Rock. Also *Solanum nigrum*, very common, called locally here black currants and much eaten. [*Solanum nigrum* does indeed bear berries like blackcurrants, and, although related to deadly nightshade and poisonous, is much eaten when the berries are very ripe, to no ill effect.] Many of the plants have been introduced from St Helena.'

Missing from his diary about Ascension, however, was any reference to Napoleon who had died less than a year before on St Helena. The *Iphigenia*, along with other ships roaming the seas to suppress the slave trade, had used Ascension Island as a victualling station and sanatorium, and St Helena lay just 750 miles from Ascension. Presumably, plants would have been traded backwards and forwards in earnest attempts to establish kitchen gardens on both barren islands. Scurvy was the black shadow which stalked all seafarers, and fresh fruit and vegetables, a known antidote, were vital.

On 6 July, he found the kitchen garden and approvingly noted that it was laid out in very good style, growing many of the culinary plants to great perfection, for example, cabbages, leeks, carrots, lettuces. 'Irresistible. On my return down the hill, I collected specimens. Returned eight p.m.' The following day, as they sailed from Ascension, his fever returned almost as severe as ever.

Nine days later, after a 'very favourable passage' they arrived at Bahia (the Bay of Salvador, Brazil). Fortunately Don was recovering quickly from his fever. The whitewashed town was a wonderful change, with its beautifully laid-out pleasure grounds, where he spotted the same sort of fig bushes seen in Africa at Accra, although known locally here as the 'umbrella tree'.

For two days Don roamed round the town, venturing up to three miles into the countryside where he found much cultivation, a very beautiful double-flowering *Oleander* and the consul's garden full of mangoes, peanuts, myrtles, irises and begonias and quantities of oranges and limes.

The ship meandered up the coast to Pernambuco, where he found a beautiful *Mimosa*, and then to Maranham, where he settled into a room, having found lodgings very difficult to get. At the same time he hired a guide who was to prove a drunkard, but his consolation was to find palms and *Amaryllis*, and to see the town at night, beautifully illuminated, like a constellation of stars. He spent his Sunday viewing local gardens, remarking that they were very rough but growing breadfruit, oranges and lemons and an 'alligator' pear (avocado).

After paddling around mangrove swamps with the inebriated guide, Don returned to his room, soaked with the heavy rain, to find an open window had let in so much wet that many of his specimens had been spoilt. Although he managed to pack most of his palms and a box of his finest shrubs, within a couple of days his fever had returned.

Two weeks later the ship anchored at Port of Spain, Trinidad, and Don took himself immediately, as per instructions from Sabine's brother back in London, to the consul's house, where he was directed to David Lockhart, the curator of the Botanic Garden in Port of Spain. It was not to be entirely as Don had hoped.

When the French Revolutionary War broke out in 1793, Britain had only four proper imperial botanic gardens, two of which were in the Caribbean. One, the empire's oldest botanical institution, was on St Vincent, established in 1765 by the maverick military governor, General Robert Melville.

In 1818, the Royal Botanic Gardens were established in Port of Spain, Trinidad, after the French-dominated island was ceded to Britain. The garden was to flourish under David Lockhart, who became an able administrator and plant hunter, but he had not managed to put the garden in order before the time of Don's visit in 1822. 'I do not believe it will ever be finished,' Don observed.

It was hardly surprising. Earlier that year, the important botanic garden at Bath, on St Vincent, had been abandoned. Its demise was not unconnected with the eccentric activities of its curator, the irascible George Caley. Plants which could be moved were shipped to the Trinidad gardens. The gardens on St Vincent had been important economically, and perhaps are best remembered as the desired destination of Bligh's troublesome breadfruit in 1787, which had a limited economic impact as spices remained the dominant product grown here. The rest, including prized nutmeg trees, were left to their fate.

Nutmeg was one of the most coveted luxuries in seventeenth-century Europe and was believed to have very powerful medicinal properties. Its price soared when the physicians in London started claiming that the spice possessed the only certain cure for the plague. In 1775, the Royal Society of Arts encouraged production of nutmeg in the West Indies. In 1831, David Lockhart must have established his own trees, as he was awarded a gold medal for the cultivation of both nutmeg and mace.

Don wisely kept his thoughts strictly for his diary. It was prudent to record that he had encountered difficulties, in case his mission was seen as a failure; in other words, he was protecting his back from being stabbed by the diehards at the Horticultural Society.

In the meantime, the gentle David Lockhart was squeezed between having to deal with the dynamic and autocratic Governor of Trinidad, Sir Ralph Woodford, and continuing to nurse the plants which had been so abruptly uprooted from St Vincent.

Sir Ralph Woodford, nicknamed *'Gouverneur Chapeau Paille'* on account of his riding around sporting a large straw hat, devoted most of the time of his administration over Trinidad, which lasted from 1813 to 1829, to the task of 'civilising' Port of Spain. He built stone houses, as opposed to the thatched houses of the previous French occupiers, constructed good roads, churches and pavements, sending out the clear message to all that the power and might of Great Britain was now here to stay.

Henry Coleridge, who arrived in 1825, wrote: 'Port of Spain is by far the finest town I saw in the West Indies, the streets wide, longer, and laid out at right angles; no house is now allowed to be built of wood, and no erection of any sort can be made except in a prescribed line. There is public water, a spacious marketplace with market house or shambles in excellent order and cleanliness.'

In 1818, Woodford purchased an estate called Paradise. A plus to this purchase was that the country now would enjoy a botanical garden. But he could not resist imprinting on it his own style. Therefore the garden was laid out with straight lines, lawns, and the imposition of order which, from his own point of view, was important. As a relatively young man, he had taken over an island culture which had been totally alien to him. Trinidad was full of different nationalities, colours and exuberance and Woodford had tried to impose a structure which he regarded as ordered and more civilised. He saw absolutely no reason why this control should not extend to a garden.

Indeed he made sure that the tropical rainforest was well cleared out of the way for the new garden, and a line of imported trees was planted as though to emphasise the control of Britain over the area. The lines of this garden can still be seen today.

The free and easy culture of the earlier rule by the French had been gradually reined in by Woodford. No longer did the 'Free Blacks' enjoy the considerable power and wealth they had built up over the last fifty years. Woodford made sure that a curfew of 9.30 p.m. was imposed on the coloured population. In the year that Don was there, a law was passed which would make it possible for a Free Black to be flogged for a minor offence on the say-so of a city magistrate, who was usually a white slave-owner.

What thoughts might have passed through Don's head, between observing the horrors of the slave ships, and the care given to rescued slaves by the British doctors in Sierra Leone, between chasing the slave ships around the coast of Brazil and now the impact of a British governor on Free Blacks. Don was a witness to history, gathering a view on slavery within a rapidly changing time.

It was all a far cry from Jamaica, for which the ship was now bound.

Fewer than twenty years before, Lady Maria Nugent, the wife of the Governor of Jamaica, revealed how she had shocked the strait-laced Jamaican colonial society by choosing to begin a ball by dancing with an elderly black slave. Knowing that her action could have precipitated a major social crisis, she retorted, 'I did what I would have done in England at any servant's hall party.' She was not entirely permissive though, and saw no reason to change her reference to slaves as 'Blackies'. On returning to England, she complained about colonial gentlemen who were noticeably lacking in manners, overindulgent and lazy, for example Lord Balcarres, whom she thought had rarely washed his hands or used a nail brush. She also complained about the heat and the mosquitoes. She pined for a Creole-style breakfast, comprising 'cassava, cakes, chocolates, coffee, tea, fruits, and all sorts of pies'.

Two weeks after leaving Port of Spain, the *Iphigenia* reached the town of Port Royal on the shores of Jamaica. Reputed in England to be not only one of the most dangerous ports in the Caribbean but also 'the wickedest city on earth', Port Royal was filled with seamen, pirates and merchants. But Don had instructions to follow, and, as in all ports of call, he had very little idea of how long the ship would remain. Immediately he went ashore to search out a Mr Higston and Dr West. The former, he found, had left for Kingston, with his affairs in a 'very sorry state'.

It was not the only calamity in Jamaica. On 11 March, there had been a severe gale in Montego Bay and, since that date and until

Don arrived on 17 October, there had been a severe drought. As he arrived, however, Don encountered torrential rain.

He took himself off to Kingston to see Dr West, but found, as in most other ports, that he could not afford to stay in the town. It would have been 'too extravagant to stay at an inn', he noted, 'when there were not many seeds close by to collect'. In addition to this dilemma, his fever had returned once more. Not surprisingly he charged the Society between three and five shillings for laundry. The discomfort of his feverish condition in the heat of the tropics was telling, but once more he managed to recover and consume a fair breakfast at the Hop Tavern, Port Royal, for 6s 6d. He also gave one shilling to 'a man for the loan of a spade'.

In company with Sabine he met 'old' Mr Wiles, who had once been superintendent of the Jamaica botanic garden. Together they climbed up to what Sabine reckoned was a mountain of 4,600 feet, and felt chilled by a temperature of 45°F.

Between Fort Henderson and Spanish Town, Don found many curious orchids, which he thought would survive greenhouse cultivation, as the night-time temperatures went as low as 40°F.

Setting sail again, on 3 November, and with Jamaica fading from view, the *Iphigenia* joined a convoy of eight merchant vessels, and a week later arrived at the Grand Cayman Islands and waited for the merchant vessels to catch up. Struggling ashore through the mosquitoes, which he thought infested the island, Don had one small triumph in finding an *Amaryllis* which he understood from the locals bore white flowers.

One week later they docked at Havana, Cuba, where Don went to see his alleged contact, one Don Antonio de la Osa, from whom the Society had had no communication for some time. After a three-mile walk searching for him, Don could make neither head nor tail of which direction to follow. De la Osa was nowhere to be found. As he realised that no one in the vicinity could either read or write English, Don quickly had his letter of introduction from Sabine translated and within a couple of days he had located the elusive Don Antonio, and found that illness had accounted for his lack of letter-writing. However, Don Antonio had an abundance of seeds and Don was willingly offered whatever he needed.

Flushed with the success of his tropical plant collecting, and sounding relieved that his health appeared to be improving, Don worked away on board the ship cataloguing, sorting and tending to his plants. They needed it. In fact they needed more than even his tender loving care could provide. He made heroic efforts to nurture his tropical finds as the ship began to plough north in heavy seas, putting his boxes down below on the lower deck. 'They will not survive long being hardly any air, if they are not chilled with cold. In any other part of the ship they will inevitably die.' On 1 December, the *Iphigenia* arrived at a freezing-cold New York.

By 9 December, the ship was still off Staten Island, anchored in the quarantine ground seven miles from New York City. 'Cold very intense', he recorded. Finally he went on a steamboat to see Dr David Hosack, who informed him about a Mr Hogg who had a small nursery, which Don confided to his diary 'will be great in a few years time'.

As the intense cold and frost held New York in its grip, Don had to hope for the best for his plants on board. He went to meet Dr Hosack who introduced him to Dr John Torrey, who published a catalogue of plants growing within 30 miles of New York. In the evening he went to a meeting of the 'literati of New York', which was held every Saturday.

He ticked off yet another botanic garden. The New York Botanical Garden was full of greenhouses from which the glass had fallen out. 'It was just like a common field', Don noted.

Despite the weather, he went with Dr Torrey to Flushing, Long Island, to a market garden belonging to a Mr Prince. Extensive it might have been, but Don saw hardly a green leaf, just some evergreens which were very common 'and an exceptionally fine plant of a variegated-leaved holly, *Ilex aquifolium*, with which Prince would not part'. Don left him with a list of plants that he would require. Digging their way under the heavy snow, however, Don and Mr Hogg did manage to uproot an *Erythronium americanum*.

On the last day of 1822, Don recorded sadly that 'I am afraid my tropical plants will die as the cold water freezes even on board the ship in a few minutes'. It was a chilly end to nearly a year in the tropics, and he was correct in his gloomy prediction. On 5 January 1823, they sailed for England, arriving on 7 February. The vast majority of his tropical plants had died, but it was not all gloom and doom. Of the many boxes he had dispatched earlier from throughout his trip, enough seeds and plants had survived to justify to the Society his journey.

As instructed in item 21 of his contract, which he had signed on 1 November 1821, he wrote immediately from 'the first port of the British Isles which you will make on your return', and also followed the instruction to 'get back with your specimens to London as soon as you can on your return'.

Although Don had shown great eagerness, courage and ability, and had lived up to his expectations in the eyes of Sabine, the task of settling down again proved far less easy, as was the case for many collectors. Aghast as he had been that so many of his tropical plants had been killed in the freezing air of New York, many of the plants he had sent home earlier had survived. Most of them were new and fascinating introductions, and some were even unknown to science.

But Don proved less willing to write up his journals. In fact, later accounts hint at a long-running dispute. Cash may have been the cause, as he had cost the Society a grand total of £143 2s 10d, along with expenses of £89 9s 9d. Most probably there was also a charge for his berth on the *Iphigenia*. The total was roughly the same as Anderson's yearly wage at the Chelsea Physic Garden. Don had risked his health and life for the same remuneration as a comfortable job in London.

While the Society might not have been best pleased with Don and his writing abilities, Don nevertheless proceeded to have the final, and extensive, say.

He became a successful writer, preparing the first supplement to J.C. Loudon's *Encyclopaedia of Gardening*, and a series on the general subjects of gardening and botany.

He settled down at 44 Bedford Place, Kensington, where in 1848, many years after he had returned from his expedition, he replied to a query from Dr Patrick Neill, he whom thirty-four years earlier had been so unimpressed by the young George Don. Dr Neill was interested in writing about Don's father. Don provided the correct details and also wrote about his brothers. Patrick appeared to have done as well as the other two, James and Charles.

Don may well have travelled to Scotland, but he never again took part in an expedition. He died in Kensington in 1856. Of the many plants he brought back, as instructed, the only plant which bears his name is *Memecylon donianum,* a tropical shrub.

In addition to the many orchids and numerous exotic plants he brought back, for growing under glass in most of Europe, he is also responsible for many of the ornamental *Allium*, types of witch hazels, *Iris foetidissima* and the wonderful *Kalmia latifolia,* or calico bush. Many plants from his difficult visit to New York, at a time when the ground was so hard that he had to hack plants out of the frozen soil, had excited the Society so much that they began to hatch another plan for a return visit to the area. This, however, was to be undertaken by another tough, young Scot, born only a year later than Don, just 25 miles from Don's native soil. His name was David Douglas.

CHAPTER 9

David Douglas (1799–1834)

THE VIRGIN FORESTS OF AMERICA

———

Young David Douglas was less interested in his school studies than in his menagerie of pets. Any pennies that he had, and these must have been as scarce as hen's teeth, were exchanged for liver to feed his pet owlets. From the age of seven, he wandered along the track to Kinnoul School, three miles from Scone, the traditional crowning place of the kings of Scotland. But he was a reluctant scholar. Far more fascinating was the natural world around him, and school life was all too often dominated by the dominie's (headmaster's) tawse (strap), which was used with frequency to drive home learning and discipline. Douglas was distracted by trapping mice, rescuing injured birds and caring for his beloved pets. Less time was devoted to books, to the rhyming singsongs of multiplication timetables and the practising to perfection of copperplate writing scratched out on a slate.

Born on 25 June 1799, at the village of Scone, a few miles north-west of Perth, he finished his early schooling at age eleven, when he became an apprentice gardener on the estate of the Earl of Mansfield at Scone Palace. His father, the local stonemason, probably fixed up the arrangement, and it is unlikely that young Douglas had much say in the matter. Luckily for him, it was a move made in heaven.

In the seven years at Scone under the strict tutorage of the head gardener, William Beattie, who disdained formal education, he was in his element; nurturing plants came as naturally to him as nurturing pets, and Beattie, a hard taskmaster, was wont to push the lad further. Although he did attend classes in Perth to learn more of the scientific and mathematical aspects of plant culture, Douglas was also absorbing a great deal of knowledge within the garden. It was a period of huge change at Scone, in the extravagant new mansion house of the Earl of Mansfield, to be called Scone Palace, and in the grounds rising up around him. To undertake this massive change, Mansfield had effected the uprooting of the entire existing village of Scone to another location, to be called New Scone. Around the newly constructed Scone Palace, landscaping on a grand scale was under way. Within a few years, and with the encouragement of Beattie, the eighteen-year-old Douglas moved to the estate of Sir Robert Preston at Valleyfield, in Fife. There he encountered a diversity of plants – both indoor and outdoor – from around the world and this opened up new horizons. Sir Robert's library was offered for his study, and he began a programme of serious self-education among these gardening and botany books.

Douglas was not unusual in his willingness to travel long distances in order to be packed into a small, drab and unprepossessing lecture hall. He tramped thirty or forty miles to attend botany lectures at the University of Glasgow, and was noticed by the Professor of Botany there, William Hooker, who was later to become a famous and hugely successful director of Kew. Hooker was a charismatic lecturer who picked out Douglas as a young man worthy of encouragement, and as an appealing character. Douglas was destined to mingle easily within new or strange social worlds, but only on his own terms: he was always very much his own man. Hooker became a father-figure, and Douglas was gradually included in Hooker's private family life, being invited to his Glasgow home, and also asked to accompany him on expeditions into the Highlands. Twenty years later, in remembering his protégé, Hooker wrote of Douglas that at that time 'his great activity, undaunted courage, singular abstemiousness and energetic zeal at once pointed him out as an individual eminently calculated to do himself credit as a scientific traveller'. Hooker recognised that Douglas' youthful, single-minded stubbornness was the ideal prerequisite for what Hooker had in mind for his next move.

Whilst Hooker's fatherly guidance was to remain with Douglas all his life, Douglas by now would have grown apart from his background through education and travel. By 1823, on the recommendation of Professor Hooker, he moved to the Horticultural Society of London, a society which was a little younger than Douglas' twenty-three years. Letters of introduction were the means by which up-and-coming gardeners, botanists and plant collectors were able both to meet and to be recommended for work. Hooker followed the usual practice of penning an introductory letter to the Horticultural Society. The Society pointed the way to good contacts in new places, so that young men wasted no time on arrival but made a beeline for the local source of knowledge. However, these letters of introduction were also only that; once 'introduced' the traveller had a very high standard to maintain in order to earn respect. In this aspect, Douglas learned to excel.

The Horticultural Society was growing each year, and Douglas landed there at an opportune time. The Society was determined to acquire, by every means possible, the exciting plants they knew existed in partially explored parts of the world. Douglas was an ideal candidate for an expedition: recommended by Hooker, young, tough, hungry for adventure and agreeable to their terms, which were thoroughly exacting. Douglas was dispatched to the east coast of North America to investigate and obtain, as gifts if possible or at least without offering payment, fruit trees and any other interesting species. He was also to investigate the latest developments in fruit-growing. Douglas was ordered to study the American apple tree for the Society, a curious instruction, as apples had in fact been introduced into North America from Europe. But the original *raison d'être* of the Horticultural Society (the Royal was not added until the middle of the nineteenth century when Prince Albert,

Japanese anemone
(*Anemone Japonica*).
Robert Fortune
described his first
glimpse of this flower
when 'it was in full
flower amongst the
graves of the natives
[Chinese] which are
round the ramparts of
Shanghai; it blooms in
November, when other
flowers have gone by'.
(From *Familiar Garden
Flowers*, fifth series,
Shirley Hibberd/
F. Edward Hulme)

Rosy Clarkia (*Clarkia pulchella*). David
Douglas found clarkias in California.
Named after American William Clark,
who in turn nicknamed them 'farewell
to spring' flowers. He had crossed the
Rockies with Merriweather Lewis in
1804–1805. (From *Familiar Garden
Flowers*, fourth series)

Pelargonium
(*Pelargonium speciosum*).
Francis Masson sent
back many varieties of
the South African
Pelargoniums, which
became the parent
plant of the thousands
of cultivated varieties
now growing.
(From *Familiar Garden
Flowers*, third series)

Nemophila (*Nemophila menziesii* or
insignis). Brilliant blue flowers with
white centres earned the apt common
name 'baby blue eyes' were discovered
by Archibald Menzies and brought back
by David Douglas from the Western
Unites States of America. (From *Familiar
Garden Flowers*, third series)

Camellia (*Camellia Japonica*). Robert Fortune brought back many varieties of camellia from the Far East. One with double red flowers bears his name; the white variety became a trademark of designer Coco Chanel. (From *Familiar Garden Flowers*, second series)

Coreopsis (*Coreopsis lanceolata*). Commonly called tickweed in its native Southern Gulf States in the USA. Thomas Drummond despatched the first seeds just before his death. Now a favourite bedding plant in Europe. (From *Familiar Garden Flowers*, second series)

Winter jasmine (*Jasminum nudiflorum*). The yellow flowers of winter jasmine discovered by Robert Fortune transformed winter gardens across Europe by bursting into flower in the darkest days of winter. (From *Familiar Garden Flowers*, second series)

Gladiolus (*Gladiolus gandavensis*). Sent back by Francis Masson from South Africa, and reputed to have been worn by Marie Antoinette as a corsage. (From *Familiar Garden Flowers*, second series)

Brassica juncea. Thomas Thomson and his childhood friend Joseph Hooker found what is commonly known now as 'oil seed rape' growing in Northern India in the late 1840s. Almost at the same time, Robert Fortune reported from China that when in flower 'the whole country seems tinged with gold, and the fragrance which fills the air, particularly after an April shower, is delightful'. (From *Kohlers Medizinal-Pflanzen* (Medicinal Plants), courtesy of the Royal Botanic Garden, Edinburgh)

Fortune's rose. Still a favourite today, Robert Fortune's double yellow rose, which was described effusively in the 1850s: 'how [can] such a beautiful flower have remained so long comparatively unknown?' In reality it is apricot yellow, suffused with carmine. (From *Curtis's Botanical Magazine*, 1787–, courtesy of the Royal Botanic Garden, Edinburgh)

Delphinium, the slender
upright larkspur raised at
the Glasgow Botanic
Garden from seeds sent
from Velasco, Texas by
Thomas Drummond in
1834, is a delicate
forerunner of the many
hybrids today. The brilliant
blue-flowered delphinium
added one more the few
very rare true blue garden
flowers available. (From
Curtis's Botanical Magazine,
courtesy of the Royal
Botanic Garden,
Edinburgh)

Lewisia. Gathered by naval
surgeon David Lyall, the
plant was found on the
south-west boundary of
British Columbia, and was
known at Kew to have the
ability to remain dormant
for several years, but 'had a
well known tenacity of life'.
This specimen flowered a
year after Lyall deposited it
at Kew in 1861. (From
Curtis's Botanical Magazine,
courtesy the Royal Botanic
Garden, Edinburgh

Phlox drummondii. Ill-fated Thomas Drummond sent back seeds of this plant just prior to his death in 1835, which was promptly named in his honour *Phlox drummondii* by Sir William Hooker of Kew. (From *Curtis's Botanical Magazine*, courtesy the Royal Botanic Garden, Edinburgh)

Magnolia. The evergreen magnolia, with its huge white waxy flowers, was gathered by John Fraser from the eastern area of the USA in the late 1700s. (From *Curtis's Botanical Magazine*, courtesy of the Royal Botanic Garden, Edinburgh)

Ranunculus lyallii. One of the largest flowered of the buttercup family, described in the *Curtis's Botanical Magazine* as 'certainly the Monarch of the genus' with 'leaves 1 foot in diameter' was often seen 'covering hundreds of acres with one huge sheet of white [flowers]'. It was discovered in the west of South Island, New Zealand and bears the name of its finder, *Ranunculus lyallii*, after naval surgeon David Lyall. (From *Curtis's Botanical Magazine*, courtesy of the Royal Botanic Garden, Edinburgh)

William Wright (1735–1810). Engraving by W.H. Lizars from a miniature by John Caldwell in
Memoir of Dr William Wright (Courtesy A. K. Bell Library, Perth)

Francis Masson (1741–1805), the first plant collector sent out from Kew to the Cape,
Canaries, Azores, West Indies, Spain and Portugal and North America. Portrait by
George Garrard, presented by W. Carruthers in 1997 to the Linnean Society.
(Courtesy of the Linnean Society of London)

Mr JOHN FRASER, F.L.S.

Lithographed for the COMPANION to CURTIS'S Botanical Magazine, from an Original Portrait in the possession of his Son, Mr John Fraser.

ABOVE.
John Fraser (1750–1811) Lithograph for the 'Companion to *Curtis's Botancial Magazine*' from an original in the possession of his son, Mr John Fraser. (Courtesy of the Royal Botanic Garden, Edinburgh)

OPPOSITE.
Archibald Menzies (1754–1842). Portrait in crayon by an unknown artist. (Courtesy of the Director and the Board of Trustees, Royal Botanic Gardens, Kew.)

ABOVE.
Thomas Drummond (c.1793–1835).
Crayon drawing by Sir Daniel Macnee.
(Courtesy of the Director and the Board
of Trustees, Royal Botanic Gardens, Kew)

RIGHT.
George Don, Sr. (1764–1814). Portrait by
John Young, taken from *Portrait Gallery of
Forfar Notables*, published in 1893 by
William Jolly & Son, Aberdeen.
(Courtesy of Forfar Library, Angus)

OPPOSITE.
David Douglas (1799–1834). Drawing by
Sir Daniel Macnee, 1828 (Courtesy of
the Director and the Board of Trustees,
Royal Botanic Gardens, Kew)

TOP LEFT.
At the Portage. Criss-crossing Canada distributing supplies and collecting furs could only be efficiently done by utilising rivers. At the points where the water-courses finished, the canoes and all supplies were carried in superhuman feats of strength across land until another river was reached. (Courtesy of Hudson's Bay Company Archives, Archives of Manitoba)

TOP RIGHT.
The Trading Store c.1840s. Stores such as these were set up across Canada by the Hudson's Bay Company. They supplied company employees with many of the essential food, firearms, snowshoes and blankets, goods also required by explorers David Douglas, Thomas Drummond and John Jeffrey. (Courtesy of Hudson's Bay Company Archives, Archives of Manitoba)

ABOVE.
York Factory, 1853. This was the arrival point for most Hudson's Bay employees on board the Hudson's Bay company ships. These brought in supplies and carried passengers such as the plant hunters sent out from Britain. (Courtesy of Hudson's Bay Company Archives, Archives of Manitoba)

TOP.

The city of Foochou and Long Bridge. (This and the fllowing illustrations are all taken from the 1847 edition of Robert Fortune's *Three Years' Wanderings in China*)

ABOVE.

The city of Ningpo – bridge of boats.

TOP LEFT.
Pagoda, Ningpo.

TOP RIGHT.
Mountain chair. Possibly a portrait of Robert Fortune himself being transported.

ABOVE.
The Shanghai river (Huangpu).

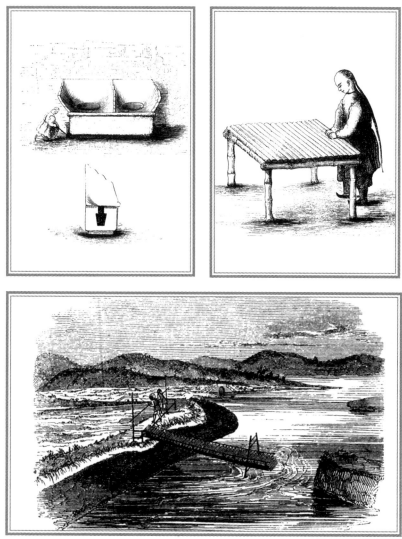

TOP.
Tombs on the island of Chusan.

MIDDLE.
Tomb of a mandarin's wife.

ABOVE.
Anemone growing on tombs.

Consort of Queen Victoria, graciously gave his support) had been to instigate a research programme for fruit breeding. Short-lived as this first purpose had been, the Society kept a sharp eye out for new and improved fruit species abroad, and were ever open to acquiring new, juicy or abundant species for the least outlay possible – they were notoriously mean – to be used for propagation, and for selling on the many plants produced.

Douglas set off from Liverpool in June 1823, bound for New York. As per his instructions, he first visited Dr David Hosack. By mid August, Douglas was down in Philadelphia looking at such treasures as the Oregon grape, or *Mahonia aquifolium*, carried back all the way from the west coast by Lewis and Clark, which even at that moment was flourishing in some American as well as European gardens. By September Douglas was in south-eastern Canada, looking as always for seeds and cuttings of fruit trees as well as wild, woody plants. Perhaps portentously, while Douglas clambered up a tree searching for mistletoe, he glanced down to see his guide running off as fast as he could, clutching Douglas' coat, money, field books and a textbook. Douglas was furious but undaunted, and simply started all over again.

He made important botanical connections while in the United States. One of the most memorable was in the autumn of 1823 when he encountered Thomas Nuttall in Philadelphia. Together the two men sought out some of the rarer plants to be found near the city, with Douglas gathering seeds for his sponsor, the Horticultural Society. He returned to London having passed his first test with flying colours, with minimum expenditure and few casualties. During the journey home, the pigeons gifted to the Society by DeWitt Clinton passed the time by fighting each other to death. Some ducks suffered agonies of seasickness but survived, and the bottles of seven-year-old cider made from the nurseryman William Cox's apples found their way to the dining tables of the Society members. The Society looked with interest upon Douglas' finds, and wasted no time in forming new plans in which Douglas was to be the central figure.

On 25 July 1824, he boarded the *William and Ann*, bound for Fort Vancouver on the Columbia River. The next place he was headed was the west coast of America. Dropping by Madeira en route, he was greatly impressed by the vines and the variety of fresh produce for sale in the markets. He and a fellow young Scot, surgeon John Scouler, followed in the tradition of many seafarers before them, such as Horatio Nelson, and purchased eight gallons of wine for around seven pounds – possibly a month's pay for both of them. In Brazil, Douglas found a Roman Catholic service so fascinating that he noted it carefully in his diary, a daring entry as religious intolerance was still endemic in the powerful societies of Britain. They called in at the remote island of Juan Fernandez, where Alexander Selkirk had been marooned, providing the basis for Daniel Defoe's novel *Robinson Crusoe*.

'As we were about to step out of the boat,' Douglas wrote in his diary,

> a man sprang out of the thicket to our astonishment and directed us into a sheltered creek. He gave me the following account of his adventures. His name, William Clark; a sailor; native of Whitechapel, London; came to the coast of Chile five years ago in a Liverpool ship called Lolland, and was there discharged. He is now in the employment of the Spaniards, who visit the island for the purpose of killing seals and wild bullocks, which are both numerous.
>
> Five of his Companions were on the opposite side, in their pursuit, and came to see him once a week; he was left to take care of the little bark and other property. When he saw the boat first he abandoned his hut and fled to the wood, thinking us to be pirates.
>
> On hearing us speak English he sprang from his place of retreat, and no language can convey the pleasure he seemed to feel. He had been there five weeks and intended to stay five more; he came from Coquimbo, in Chile. His clothing was one pair of blue woollen trousers, a flannel and a cotton shirt, and a hat, but he chose to go bareheaded; he had no coat. The surgeon and I gave

him as much as could be spared from our small stock, for which he expressed many thanks. His little hut was made of turf and stones thatched with the Straw of a wild oat. In one corner lay a bunch of straw and his blanket; a log of wood to sit on was all the furniture; the only cooking utensil was a common cast-iron pot with a wooden bottom, which he had sunk a few inches in the floor – and placed the fire round the sides! He longed to taste roast beef (having had none for seven years) and one day tried to indulge with a little baked, as he termed it; but in the baking the bottom gave way, as might reasonably be expected; so poor Clark could not effect the new mode of cooking. I told him under his circumstances roasting beef was an easier task than boiling. He is a man of some information; his library amounted to seventeen volumes – Bible and Book of Common Prayer, which he had to keep in a secret place when his Spanish friends were there; and an odd volume of *Tales of My Landlord* and *Old Mortality*. He had . . . [a] fine bound copy of Crusoe's adventures, who himself was the latest and most complete edition.

This last perhaps Douglas noted with an eye on his own diary's publication. The novel *Robinson Crusoe* had been published over a century before Douglas visited Juan Fernandez. Like so many of the places and people he was to come across during his travels round to the other side of the globe, there was a strange element of familiarity, a brush with homespun Scottishness.

At the Galapagos Islands, he was frustrated by the brevity of the opportunity to go ashore. He recognised, having more time on his hands there than Archibald Menzies, and Darwin many years before, the sheer volume of plants and animals not accurately studied or catalogued. Many questions arose in his probing brain. Finally, the expedition reached the Columbia River, their destination, in April 1825. He began to collect at once. The rewards were immediate and numerous. Even though first Menzies and then Lewis and Clark had collected plants in the area, they had found only the obvious ones. They had been more intent on covering the miles, whereas Douglas was in the field with more freedom to wander. He was

finding curious plants that proved to be new to science, and as
he travelled up the Columbia River, the novelties became more
frequent. Because he was working for the Horticultural Society,
Douglas observed flowering and fruiting material, and often had
to return to the area at a different time of the year to gather seeds.
Not only did this require a feat of memory, he also had gained
from the intensive instruction in reading compasses and various
instruments that proved vital in pinpointing his position. This
had been arranged by Joseph Sabine the lawyer who was the
Administrative Secretary of the Society in London, whose naval
brother, Captain Edward Sabine, was ideal for the task. Instead
of returning to London in 1826, as instructed by Joseph Sabine,
Douglas took a huge risk and decided to stay in America. There
were too many new plants yet to be collected.

Douglas was the first man to be posted at Fort Vancouver to
pursue anything other than hunting or trading. He was a rare
commodity, a non-exploiter of the land. Others might have been
inclined to plunder rather than adapt, erasing nature as they went.
Douglas was merely plucking minuscule treasures in the forms of
seeds, the occasional bird specimen and bulbs.

As Douglas' ship had sailed up the Columbia River, he had
written of the trees, the Douglas firs which were to carry his name,
and *Gaultheria shallon*, a pretty purple-berried evergreen, about
which he was so thrilled he could hardly see anything beyond
it. Dividing the Columbia River, opposite Fort Vancouver, was
Menzies Island, named after that illustrious Scottish naturalist,
whom Douglas had already met at the elderly man's home in
London. Douglas found on this island a new form of *Myosotis*, a
bright-yellow forget-me-not which he and the Glaswegian surgeon
John Scouler promptly named after Hooker. Douglas was much
pleased with this, especially as it was in flower in late September,
when flowers are scarce. He was eventually to dispatch about
eighteen varieties of lupin, coloured mainly blues and purples but
including *Lupinus sulphurous*, a shrub-like plant with spurs of sweet-
smelling yellow flowers. He also sent back *Phalangium quamash*

(camas lilies) with their spikes of hyacinth-blue flowers, whose bulbous roots, he had observed, formed a great part of the natives' food. After cooking these bulbs, Douglas described the taste as much like a baked pear. Another bulbous meal was cooked up from the corms of *Fritillaria pudica,* a delightful fritillary with nodding heads of yellow flowers which gradually turn orange to brick-red with age. *Pudicum* means bashful, an apt name as when the flower emerges there is little hint of the glories to come. This he sent home in a jar of sand. On his arrival at Fort Vancouver, Douglas met Francis Ermatinger, and a namesake, James Douglas, destined to be Governor of Vancouver Island and British Columbia. Immediately, Douglas was swept up into the delicate and time-consuming art of dealing with the local Indians.

On Saturday, June 17th. – my guide did not arrive from the camp until 8 a.m. And as I was uncertain if he would come that day, the horses were not brought in from the meadow, nor my provisions put up. Considerable time was taken up explaining to him the nature of my journey, which was done in the following way: I told Mr Black in English my intended route, who translated it to his Canadian interpreter, and this person communicated it to the Indian in the Kyuuse language, to which tribe he belongs.

As a proof of the fickle disposition and keenness of bargain making in these people, he made without delay strict inquiry what he should get for his trouble. This being soon settled then came the smaller list of present wants, beginning, as his family had been starving for the last two months, and he going just at the commencement of the salmon season, by asking Mr Black to allow them something to eat should they call, which was promised.

Afterwards a pair of shoes and, as his leggings were much worn, leather to make new ones was necessary; a scalping knife, a small piece of tobacco, and a strip of red coarse cloth to make an ornamental cap. This occupied two hours and was sealed by volumes of smoke from a large stone pipe. Mr B. kindly offered to send a boy twelve years of age, the son of the interpreter, who speaks the language fluently, with me, which I gladly accepted. As he spoke a little French, I would be the better able to make known my wants to my guide. I had provided for me three excellent horses for carrying my paper, blanket, and provisions, which was equally divided, and as I choose to walk except on bad places of the road or crossing the creeks, I placed a little more on my horse.

A further insight into the complex story of dealing with the native peoples concerned alcohol, an aspect which Douglas viewed with increasing dismay. He noted down that 'In the evening I gave the two chiefs a dram of well-watered rum, which pernicious liquor they will make any sacrifice to obtain. I found an exception in my guide Tha-a-muxi; he would not taste any. I enquired the reason, when he informed me with much merriment that some years since he got drunk and became very quarrelsome in his village; so much so that the young men had to bind his hands and feet which he looked upon as a great affront. He has not tasted any since.' The story of the introduction of potent and powerful alcohol into native society by invasive Europeans in the remote areas of the North Americas is, with the benefit of hindsight, a thoroughly appalling saga. Douglas was to observe the effects at close range. Rum became elevated to the major currency of the fur trade, with vast quantities being bartered for pelts. The traders discovered the high value of their rum amongst the native peoples and exploited their desire and rapid dependency for astonishing financial gain.

While Douglas was in the country, the Hudson's Bay Company shrewdly found this Achilles heel in the Native Americans' bargaining tactics and rolled overland around 50,000 gallons of strong rum or brandy as payment. This was a staggering quantity, as when diluted, it would create at least a quarter of a million gallons or 1.1 million litres when the native population in total numbered only about 120,000.

Watering the spirit down led to the birth of the expression 'firewater': the local recipients would spit a mouthful on the fire before accepting their side of the bargain. If the fire spurted with flame, the spirit was strong, if it partly quenched a fire, then it was well watered down, and was rejected as an adequate payment. All the fur trading companies accepted that cheating occurred, usually at the point of distribution from the company officials to the native traders. One of the most devious ways of gaining even more profit was to imitate rum by adding a few drops of iodine to raw gin, a cheaper liquor. Many employees of the Hudson's Bay Company pleaded with their seniors in command in London to put an end to this pernicious trade, which led to mass addiction and the destruction of many indigenous ways of life along with families and whole communities. Their pleas were to no avail.

Food was a topic of constant interest to all Hudson's Bay employees and to Douglas as well. He kept a ready supply of pemmican and a little biscuit, sugar and tea. Tea was one of the most important provisions for Douglas. Tobacco he might use for exchange and barter, game he might shoot and give away, items of clothing he would gladly exchange for a guide or much sought-after seeds, but tea he never appears to have traded. Pemmican was the essential all-purpose food for everyone from the lowliest trapper, canoe man and even traveller to the high table – in dire necessity only – of the resident factor. Originating from amongst the native North Americans, it resembled dried animal food, and was concocted from dried buffalo meat, the fat of any available animal, and dried berries from one of the many available varieties of the *Vaccinium* family which produce many tart little berries packed with vitamin C, such as blueberries and cranberries. The proportions were simple: one pound of dried buffalo meat to sixteen pounds of berries. The method was equally simple: the hide of the buffalo was sewn into a sack which would contain about ninety pounds of pemmican when complete. This sack was half filled with pounded and pulverised meat, and topped up with dried berries. Finally the boiling fat was tipped in, the contents stirred, the hide sewn up and the contents left to set. This basic food was carried across thousands of miles, and eaten raw, sliced up and fried, or made into a thick soup. It was reputed to last for ever. Some remembered pemmican with affection, others were less sure. H.M. Robinson memorably described his impression in *The Great Fur Land* as follows: 'Take the scrapings from the driest outside corner of a very stale piece of cold roast beef, add to it lumps of rancid fat, then garnish all with long human hairs and short hairs of dogs and oxen and you will have a fair imitation of common Pemmican.'

If Douglas was a moderate shot before he arrived, his accuracy improved dramatically once he landed. If he didn't want to eat pemmican, he depended on shooting to supply food for the pot and he also needed to be able to defend himself. Sometimes opportunities for gun use came one on top of the other. Real-life plant collecting in the American west required ingenuity and the sugar pines which Douglas had heard about presented him with a tall order. All the cones which still contained seeds were atop trees a hundred or so metres high. He solved the problem by shooting through the twigs holding the cones, a considerable feat of accuracy. However, along with the seeds, this produced an unexpected and unwelcome surprise as out of the forest came several local 'warriors, painted with red earth, armed with bows, arrows, spears, bone and flint knives, and seemed to me to be anything but friendly'. Drawing out both his pistol and gun, he pointed them at the men, and levelled his gaze for at least ten minutes, a ploy which he had learned from a senior Hudson's Bay manager, who had once spent an entire day staring down a large body of threatening Native Americans. His cool nerve paid off, and when at length one of the men made a sign for tobacco, Douglas signalled that he would give them some when they went off and picked up some cones for him. As soon as they were out of sight, he ran for safety to his camp where he dismissed his Native American servant, suspecting he also might betray his presence. Douglas then spent the night keeping awake in case of attack, and filling in the hours by writing up his diary. He was getting to grips with travel in the wilderness.

On 20 March 1827 he left with the Hudson's Bay Express on the first leg of his trip home which would take him from the relative comfort of Fort Vancouver, over the Rockies and to the shores of the Hudson Bay to await a ship home. The flat-bottomed boats of the Express spent their existence almost constantly on the move, ferrying supplies and men from fort to fort, from literally one side of Canada to the other. It was the accepted way of travelling with one boat being used for passengers whilst another carried supplies and kept alert for attack. The men who manned the boats were known as voyageurs. They often had native as well as French Canadian blood and became legends in their own lifetimes, vying with one another in speed and physical endurance. In their birch bark canoes they would travel for twelve to fourteen hours a day transporting furs and goods, and singing their songs to maintain the rhythm of the paddle strokes.

Having slogged up the Columbia River as far as the voyageurs of the Hudson's Bay Express were due to go, they now entered what was for Douglas new country. His party would now proceed on foot and in a few days reached Fort Assiniboine. As an aside, he picked out what he thought to be the highest peak in the area, which he estimated to be around 16,000 feet, and climbed it. His fascination with mountains and, later, with volcanoes, and his prowess at mountaineering, would have been astounding if it was not seen alongside the sheer toughness of the local Hudson's Bay men. They were used to living for months totally alone, constantly on edge and aware of ever present dangers, from marauding grizzly bears to hostile Native Americans.

At this point, Douglas became frustrated by the difficulties of searching for plants in such a country. He therefore turned his attention to the very necessary task of hunting large game, which now became a daily necessity in order to feed the party. Arriving at Fort Assiniboine, Douglas discovered that some of his precious seeds had arrived before him via the Express and were now travelling with another young Scottish botanist, Thomas Drummond. Horror that his collections had been taken without his knowledge made Douglas rush off ahead of the party to the next fort at Edmonton. His apprehension turned rapidly to warmth and gratitude when he caught up with the party; he warmed to Drummond instantly, especially when he found that the packages had been soaking wet, and would have rotted if Drummond had not taken prompt action. Only Douglas' shirts had prevented total ruin, by absorbing some of the water. Thomas Drummond hailed from near Forfar, only twenty miles from Douglas' birthplace. Douglas was greatly impressed by Drummond's own collection, and this meeting took the edge off his disappointment that his much-looked-forward to trip down the Red River might be curtailed owing to lack of time. Drummond and Douglas took separate ways here, but were to meet again to sail home. Douglas proceeded on with the main party, still hoping there might be time to fit in his Red River trip, while Drummond detoured to explore other areas.

There was yet more consolation to be had. He acquired a Calumet eagle as a pet. Catching eagles was a rite of passage for young warriors of the area, as the birds were well known for their ferocity. The usual method was to dig out a pit in which a man could hide under a canopy of logs and brushwood. Tempting bait was then left for the eagle and when it swooped, the man, arms protected with leather gauntlets, would grab it and hold on for dear life. This one, however, had been taken from its almost inaccessible cliff-top nest, and Douglas noted nonchalantly that he had some difficulty in putting him in a cage, but succeeded. Such was his gift with animals that the eagle was soon travelling on his shoulder, attached by a leather thong. Together with his eagle, Douglas' party continued the march, and then canoed down the Saskatchewan River to Fort Carlton House. He was now three-quarters of the way to York Factory and home, but there were several surprises in store yet:

> On Wednesday at sunrise five large buffalo bulls were seen standing on a sandbank of the river. Mr Harriott, who is a skilful hunter, departed and killed two, and wounded two more; all would have fallen had not some of the others imprudently given them the wind, that is on the wind side.
>
> Fifty miles further down a herd was seen, and plans laid for hunting in the morning. Some deer were killed this evening and some of the Prong-horned antelope of the plains. This little animal is remarkably curious in his disposition; on seeing you he will at

first give three or four jumps from you, return slowly up to within a hundred or a hundred and fifty yards, stand, give a snort, and again jump backwards. A red handkerchief or white shirt, in fact, any vivid colour – will attract them out, and hunters crawl to them on all fours, raising the back like a quadruped walking, and readily kill them.

Drummond's and Douglas' paths criss-crossed over the vast land mass of Canada, under the shelter, guidance and protection of the Hudson's Bay Company. Through Drummond, Douglas met the Arctic explorer Sir John Franklin who promptly offered him a passage in his canoe across Lake Winnipeg. For a man intent on exploring the toughest terrain and seas of the world, Franklin had a very soft side, and would help out anyone whom he thought deserving. As this saved Douglas considerable time he gladly accepted. On 10 August 1827 Douglas reached the Hudson Bay, but sadly his pet eagle, which he had carried 2,000 miles, died from being accidentally strangled on the cord that restrained him just before they reached their destination. His owner was heartbroken. At the Hudson Bay, Douglas was 'kindly received' by the chief factor, John McTavish, who thoughtfully provided new clothes for the worn-out traveller. Douglas' own words best sum up his feelings at the end of his incredible trip: 'Here ended my labours, and I may be allowed to state, that when the natural difficulties of passing through a new country are taken into view, with the hostile disposition of the native tribes, and the almost insuperable inconveniences that daily occur, I have great reason to consider myself a highly favoured individual. All that my feeble exertions may have affected, only stimulate me to fresh exertions.'

The route across the continent from the Pacific to Hudson Bay may have been familiar to the fur trappers but Douglas was the first outsider to make the trip successfully. By his own calculations he had travelled nearly 10,000 miles on foot, horseback or by canoe. Yet he had one final adventure left to make before he sailed for England. Meeting up again with his fellow plant collector Drummond, they and a party of others rowed out to visit the ship anchored in the bay. A tremendous storm blew up and drove their rowing boat out some seventy miles into the bay and out of sight of land. Given up for lost, the party had to wait in the boat a further two days and nights before the storm dropped and they could row back to land. Douglas, worn out from his epic travels, was horribly ill and had lost the power of his limbs to the extent that he spent the whole of the passage home in his quarters unable to move or even write up his journal. His companion, Thomas Drummond, claimed to suffer hardly a jot. Off he went for a five-mile walk the day after their arrival home, to stretch his legs.

Douglas' arrival in London, on 11 October 1827, was an exciting time for the Horticultural Society, but for Douglas, ill and tired after his momentous finale adrift with Drummond, the adulation was too much to cope with. Too ill even to stagger as far as the Linnean Society to present his carefully prepared report on his treasured sugar pine, the paper had to be read by Sabine on his behalf. That he had survived at all, and come home to be able to tell the tale, was relief enough for the Society. Some nervousness had been evident for a long time about the chances of Douglas returning alive at all; they were even given voice by the President of the Horticultural Society, Thomas Andrew Knight, who had penned a doom-laden letter, prophesying that: 'Our collector proposes, when he has sent all he can home by a ship, to march across the continent of America to the country of the United States on this side, and to collect what plants and seeds he can in his journey: but it is probable that he will perish in the attempt. Mr Sabine says that if he escapes, he will soon perish in some other hardy enterprise or other. It is really lamentable that so fine a fellow should be sacrificed.'

In 1825 and 1826 Douglas had marched, climbed, canoed and staggered across a total of over 6,000 miles in rough territory, far from civilisation, showing courage, tenacity and an acute sense of observation as well as a love of science and a passion for nature. He was the first European to climb the northern Rocky Mountains and in so doing named a number of them, including Mount Hooker, after his professor at Glasgow University. In 1827 he travelled across Canada, passing through the settlement of Scots at Red River, established by Lord Selkirk in 1812. He eventually reached Hudson Bay and a ship home. Hooker might well have been the recipient of an honour in having a mountain named after him, but he was also the recipient of growing complaints about Douglas in London. Douglas reacted to the polite drawing rooms of London with surly contempt. He refused to clean himself up, and took to wandering around as though he was in the backwoods of North America. This did not go down well at all, and Hooker, who had championed Douglas as a package of intelligence, physical prowess and botanical knowledge, was on the receiving end. Douglas' determination and underlying stubbornness, the very characteristics which Hooker had recognised would carry Douglas on, were now revealing themselves in a negative light. Hooker took charge. He penned letters to Douglas and to the Society. He persuaded Douglas to come north to Glasgow, where he effected a transformation by arranging for Douglas to sit for his portrait. He recognised that Douglas was furious to find that animal skins he had sent, risking his life for their capture, were rotting in a corner of the Society. The Society was, in turn, trumpeting the cost of the expedition, which, at £400 including a mere £66 for Douglas' food expenses, was roughly the same as the return on just one plant – *Ribes sanguineum*, the flowering currant – out of the 210 species which he had brought back. And yet, Douglas was paid less than the doorman at the Society's front door.

While Hooker was delighted to see Douglas in person, Douglas for his part was astonished at the transformation which had taken place at the Glasgow Botanic Gardens under Hooker's encouraging direction. Douglas' seeds and the resultant specimens proved extremely popular when they reached their destination, with plants of the Noble fir being distributed to Fellows of the Horticultural Society for as much as fifteen to twenty guineas each. Despite these positive aspects, Douglas returned to London and passed another miserable few months during the late autumn of 1828, only brightened by meeting characters like Archibald Menzies, whose trip with Vancouver, and the descriptions he had brought back of the area, had lit the touch-paper of interest in the North American west coast.

Hooker realised that Douglas would be happiest sent on a return expedition, and a combination of Hooker pressing this to the Society, and Douglas being asked to consult on the boundary between the fledgling United States and Canada, boosted his morale. In the late summer of 1829, Douglas went north to see his mother in Scone, visiting Hooker and buying a small Scotch terrier, Billy, to take with him on the next trip. He was destined for California. He departed on board a Hudson's Bay ship, the *Eagle* – at least the Bay men had been impressed by him – and landed first at Fort Vancouver. Making plans, Douglas had wanted to search once more for the elusive sugar pines, and then, travelling south, link this to his trip down to California.

North of Santa Cruz, Douglas came upon 'the great beauty of Californian vegetation', the *Sequoia sempervirens* in what is now the Redwoods State Park. He then travelled north to San Francisco and botanised in the Mount Diablo region. While waiting for a ship, he busied himself cataloguing his finds in preparation for their journey home. A golden interlude in his life at this point was a meeting and subsequent friendship with Dr Thomas Coulter, an Irish physician who had worked for the Real del Monte Company in Mexico, and was now working as a botanist for Professor Candolle of Geneva. Coulter was one of the first to botanise seriously and systematically in California. The two men got on famously, with a lack of rivalry that allowed Douglas to obtain plants from Colorado and Gila, which he carefully dispatched just for Hooker. Much as he loved the kinder climate, his eyesight was now deteriorating badly. Snow-blindness and now brilliant sunshine meant that he could not even read what he had written. Uprisings against the Mexican authorities also meant he was virtually press-ganged into a 'Company of Foreigners' to keep the peace. The routine of guard duty did not at all please a man of Douglas' disposition, especially given the brevity of the springtime for collecting plants, and he managed to extricate himself.

Back in Monterey, Douglas packed two separate consignments, one for his favourite old master Hooker and the other for the Horticultural Society. His discoveries from California surpassed his collections from the Columbia River area and the Horticultural Society were swamped for a time with sorting and cataloguing the seeds and plants. These included many varieties of *Calochortus*, or fairy lanterns, now the mainstay of every florist in Europe, along with lupins and heliotropes, Monterey pines, and the tasselled shrub, *Garrya elliptica*. While it was easy to go from British Oregon to Mexican California, it proved difficult to go back the other way. During Douglas' extended stay in California he collected something over 500 new species of plants, many of them destined to become important garden herbs, trees and shrubs. He also gathered here, as elsewhere, many different kinds of mosses, a group of particular interest to Hooker.

In August 1832 he had visited Honolulu, and while there heard of the resignation of one of his mentors, Joseph Sabine, from the Horticultural Society, which had sunk into debt. Douglas immediately resigned in protest, and from then on was a freelance collector, sending everything he could only to Hooker. By October he had returned to the Columbia River and Fort Vancouver. He formed a grand plan. Free from the constraints of an employer, he was able to make his own decisions, but also his own mistakes. He had to locate plants and seeds and make sure he could earn enough from their sale to support himself. His elderly mother also required support: his father had died and Douglas and his brother, in the accepted practice of the times, would have been expected to support their elderly parent. With this in mind, he made a simple decision: he decided to walk home.

It was not altogether a decision made on the spur of the moment. Back in London he had formulated a trip which would take him through Alaska, then Siberia – 'people tell me it is like a rat trap, no difficulty in entering, not so easy to find egress' – then across

Russia, and home. It was, after all, a time when men thought little of spending years at sea away from home, and armies marched across Europe. Few traversed by wagon or horse. Walking was a method of travel available to all, and it had a great advantage: it was cheap. For a newly independent collector, self-employed and fancy-free, it was all too obvious.

He left with the Hudson's Bay Express to Fort Okanogan, then went by horse up the Okanogan Valley to the nearby lake of the same name, and then on over the trail to Fort Kamloops. His trip seemed dogged by bad luck. Just before leaving, fever struck and 'only three individuals out of one hundred and forty escaped it, and I was one of that small number'. Not that anything so slight could put him off his mission. Even his physical strength was crumbling. His right eye was entirely blind and the left prone to double vision and blurring. He complained of using his purple eye-glasses, 'most reluctantly, as every object, plant and all, is thus rendered of the same colour'. Added to this were repeated, debilitating attacks of rheumatism, unusual in a man of thirty-four, which could leave him immobilised for up to a week at a time.

The chief trader at Fort Kamloops was another Scotsman, Samuel Black, an enormous and powerful man. During the course of the evening Douglas expressed to him his opinion that the Hudson's Bay Company was a mercenary enterprise, caring nothing for the lives of its employees in the pursuit of profits, and that there wasn't a trader in it with a soul above a beaver-skin. As was his manner, Douglas was gruff, and to the point. He had little truck with niceties, and was all too prone to say precisely what was on his mind. His scientific background and growing environmental awareness – almost a century before another Scotsman, John Muir, was to cajole the American President into saving the very same type of forests with which Douglas was familiar – was wildly at odds with the company's policy of putting profit before any environmental side-effects which might result. Black's brother was the editor of the *London Morning Chronicle* and he himself was by no means uneducated or unsympathetic to the sciences. He was,

DAVID DOUGLAS (1799–1834) 201

however, up in a second responding in a similarly insulting way to Douglas and demanding satisfaction. This challenge to duel was accepted but postponed till the following morning as it was pitch dark outside.

At first light, Samuel Black was up and demanding that Douglas rise and fight. Douglas declined, realising discretion was the better part of valour, and that he had spoken too hastily. This incident was to dog the stories of Douglas in future reports. Putting the incident behind him, he set off from Fort Kamloops up the Fraser River towards Fort Alexander, walking until he reached the Quesnel River, then boarding the Hudson's Bay boats, used by the company for transporting furs, supplies and men. On 13 June at Fort George Canyon disaster struck. They capsized and Douglas was trapped in the whirlpools and swept downstream for an hour and forty minutes. Although he eventually managed to retrieve his instruments, charts and some of his barometric readings, he lost his clothing, seeds, plants, and his precious diary. Miraculously, Billy the terrier survived.

Battered, soaked, miserable, and with his dreams evaporating, he slowly trekked back, saying farewell to his idea of walking home. He had covered but the tiniest fraction of the journey and lost virtually all his possessions. Another canoe was found, and they retraced their journey, descended the Fraser, reached Fort Okanogan and then carried on down the Columbia to Fort Walla Walla. Summoning up all his strength, Douglas stopped off to visit the Blue Mountains in order to go botanising and recover some of the lost plants. He returned to Fort Vancouver by August, utterly despondent and nursing a fragile spirit and crumbling body.

He remained in Oregon only a few weeks and then sailed for Hawaii in mid October, arriving there just before Christmas 1833. Douglas' winter visits to Hawaii were a routine event as gathering plants on the floristically rich islands was as rewarding botanically as it was in the rich forests of Oregon. He also took barometric measurements of several of the mountains, climbing Mauna Loa and Mauna Kea, one of the first Europeans to climb both. He stayed into July, the idea being that he would return shortly to London. In the summer of 1834, Douglas was waiting for a passage back to Britain when he met John Diell, a chaplain from the seaman's mission. Diell was keen to see something of the spectacular volcanic mountains of Hawaii and the two men struck up a friendship. Douglas agreed to show him the sights of Kilauea while he was waiting for news of his ship.

In July, the two men set off, but in the first of a series of odd happenings, they became separated when Diell made a detour to visit Molokai. Douglas did not turn back at that point, but continued on his way with John Diell's manservant and the faithful terrier Billy, his constant companion. The party had intended to continue walking to Hilo, a distance of some ninety miles, but Diell's servant also dropped out having developed some form of 'lameness'. Douglas went on alone with Billy as his only companion. On 12 July, he came to the huts of a cattle hunter, Edward (Ned) Gurney, who made an opportunistic living by trapping wild cattle in pits on the hillside and selling the meat, hides and tallow to passing ships. Gurney was an ex-convict from New South Wales, deported there from London in 1819 for theft, and locally rumoured to have escaped to Hawaii in 1822.

After breakfast Douglas asked Gurney for directions and the cattle hunter accompanied him a few miles up the track, warning him about the existence of three cattle pits nearby. Gurney was later to claim he had turned homewards around 10 a.m. A couple of hours later, two natives, perhaps alerted by the shrill yelping of Billy the dog, went over to the pits. To their horror, they looked in at a body, bloody, trampled, broken and quite still, lying at the foot of the pit. Still very much alive was a trapped bull. Gurney was rushed to the scene, where he shot the bull and rescued the mangled body. He recognised it to be that of Douglas, who was clearly dead. Close by was Billy, still guarding a bundle belonging to Douglas, which appeared to have been left on the path to Hilo. Later, Gurney claimed that he had followed Douglas' footprints from one pit to the other, and finally to the third pit, where he must have slipped and fallen in, possibly when peering down inside. Once in the pit, he had been gored and trampled to death by the trapped, possibly injured and infuriated bull.

Gurney was probably the last man to see Douglas alive. Rumours began to circulate that he had been murdered. Few would have looked kindly upon Gurney, whose past would have been enough to raise suspicion. Gurney had ordered the body to be wrapped in hide and deputed a party of local Hawaiians to carry it the twenty-seven miles to Hilo, finishing the journey by sea canoe, where Diell and Goodrich, an American missionary friend of Douglas', were both waiting for Douglas' arrival as previously arranged. To their horror they found that their friend was a mutilated and, as a result of the summer heat, rapidly decomposing corpse. After the shock of Douglas' death, the resident missionaries, Lyman and Goodrich, began to question the manner of his demise after they were approached by the local carpenter who was constructing the coffin and dealing with the body. They realised that his purse, which had contained money to pay for his guides, and possibly enough to pay his passage home, was missing. This was a perfectly credible pronouncement as Douglas had written in his journal about paying for his guides.

Sarah Joiner Lyman of Hilo, the wife of missionary David Lyman, recorded the day:

> July 14th 1834. This has been one of the most gloomy days I ever witnessed . . . Mournful to relate Mr Douglas is no more . . . just as Mr Diell was about to go down to the beach to meet Mr Douglas, we were informed that his corpse was at the water's edge in a canoe . . . his clothes are sadly torn and his body dreadfully mangled. Ten gashes on his head . . . A carpenter was engaged to make a coffin and a foreigner to dig his grave under a breadfruit tree in Mr Goodrich's garden. Whilst engaged in digging the thought occurred to him that Mr Douglas was murdered. He suggested it to Mr Goodrich and Mr Diell. Their suspicions were at once excited. They left digging the grave and concluded to preserve the body in salt and send it to Honolulu that it might be

more satisfactorily determined how the wounds were inflicted . . . the whole is involved in mystery.

The 'foreigner', an American bullock-hunter named Charles Hall, was the first to raise doubts. Experienced in animal trapping, perhaps he was doubtful if Douglas' injuries were consistent with the death he was supposed to have suffered. Hall was an experienced hunter of Hawaii's wild cattle, which had increased in number greatly since Vancouver landed the first animals on the island.

Hall was later to contend that Gurney and Douglas had quarrelled, Gurney killing the botanist and disposing of the body in a wild bull pit – a mere 300 feet from Gurney's house – to cover up the evidence of injury by ensuring that the bullock would damage the body. Hall was to emphasise that, in his opinion, the blunt and battered horns of an elderly bull such as the one trapped would have been incapable of inflicting such injuries.

By the time Douglas' body was examined by four doctors who had been hastily assembled by the British Consul, it was in 'a most offensive state' and the doctors found nothing to suggest foul play. Hall and another white man were sent to the bullock pits to investigate them. They heard several reports from locals in the area which cast doubt on Gurney's story, one of the most compelling being that Douglas was said to have been carrying 'a large purse of money' which had subsequently vanished. The footprints described by Gurney had been washed away by heavy rainfall and there was no other evidence relating to the death.

The funeral was held on 4 August in the church of Kawaiaho, and was attended by the foreign residents of Honolulu and officers of the *Challenger*, the ship whose doctors had carried out the post-mortem. The service was read by one of the officers and the body interred in the burial grounds of the church. The suspicions that Douglas had been murdered persisted after the funeral, and were in no way allayed by the findings of the post-mortem. Ned Gurney was the chief suspect, but doubts were also held about John Diell's servant, who had disappeared before Douglas' death. It was proposed that

Douglas' relations with the native Hawaiians were not as good as they had been, mainly due to the botanist's increasing irritability, resulting from his deteriorating health. Was his death the result of a quarrel? Then there was the problem of the missing purse. No one knew for sure if such a purse had existed although a few dollars were found on Douglas' body after his death. The chronometer he carried in his breast pocket was smashed yet the case in which it was held was intact, strange as logically it should have been the other way round. Had someone smashed the chronometer and then replaced the protective case? Some doubted that, bad eyesight notwithstanding, anyone could have fallen into a pit which had already had its cover broken by a bullock.

Many of the Hawaiians themselves believed that Douglas had been murdered, and a report was published in the *Hilo Tribune* sixty-two years later, in 1896, claiming that a native hunter named Bolabola knew for certain that this was the case. A noted hunter and trailer, now over seventy years of age and familiar with every inch of ground around the Mauna Kea slopes, the area in which Douglas met his death, Bolabola's tale was recounted in the customary, slightly breathless style of local newspaper reporting:

> It was noticed that he [Bolabola] lowered his voice, with now and then the older men nodding in approval. Although he had been only ten years old at the time, and all what he had to say was simply hearsay, he insisted that Douglas had been murdered . . . we all felt so at the time, but were afraid to say so and only whispered it among ourselves. And when my father and the old Kaline [another noted Hawaiian] died, they both repeated the story to me.

Clearly Bolabola had been convinced by his elders over the years that Douglas had been murdered.

Other facts are more consistent with the accident theory. It transpired that people had been known to fall into cattle pits. Also, Douglas' dog was still guarding his master's bundle by the path, and if there had been a struggle it was felt that the dog would have been involved. Douglas' eyesight was not good. Curious as ever, he could easily have crept too close to the edge, and, in trying to peer into the pit, he could have simply slipped and fallen in.

In 1989 Joseph Theroux, a resident of Hilo, wrote an article called the 'Mauna Kea Killing' in which he confirms his conviction that Douglas was murdered by Gurney, even turning his theory into a novel based on his research on the life of Ned Gurney. However, until the body is exhumed and modern forensic methods are brought to bear, and the gash on the skull decided on one way or the other, we will never know the answer. As it stands, the mystery is unsolved.

The British Consul in Hawaii, Richard Charlton, now had the gloomy task of selling Douglas' clothing and forwarding his collection of books, papers and instruments to the secretary of the Horticultural Society. This task was fraught with problems, due to the distances involved, and lack of clarity as to who should be informed. Douglas must have revealed that he was now working for himself, as a freelance collector. Eventually, William Hooker, the man who was his mentor, father-figure and a personal friend, was to hear of Douglas' death in a particularly roundabout fashion.

> The first knowledge of his [David Douglas'] decease, which it reached one of the members of his family in this country, was in a peculiarly abrupt and painful manner. It was seen in a number of the *Liverpool Mercury*, by his brother Mr John Douglas, when looking for the announcement of the marriage of a near relative. He immediately set out for Glasgow to communicate the unwelcome tidings to me; and in a few days they were confirmed on more unquestionable authority, by letter from Richard Charlton Her Majesty's Consul at the Sandwich Islands to James Bandinel Esq.

James Bandinel was in the Foreign Office and stationed on the islands. Hooker was to take over some of Douglas' collection, including several preserved birds. Billy the dog set sail for England where he spent the rest of his life with Mr Bandinel.

The trip to the Pacific North-west by sea via Madeira, Rio, Cape Horn and the Galapagos was remarkable in itself, yet Douglas accomplished this twice. The overland trips around the Columbia River undertaken by Douglas have given an enormous insight into the lives of the earliest Europeans to settle and trade in that area. Sometimes in reading Douglas' journal it is easy to take for granted the difficulties of travel at that time. Douglas seems to have faced up to all of this with a remarkable sangfroid. In his contacts with the Native Americans, Douglas adopted a pragmatic approach insofar as the natives were often the best people to help him achieve his objective of collecting seeds and plant specimens. The fact that Douglas generally got on well with the people whose country he was travelling through bears testimony to his negotiating skills. Like so many others, Douglas was an ordinary man who achieved extraordinary things in his lifetime. And like so many ordinary men, he certainly wasn't perfect. There is no doubt that Douglas could be a difficult and stubborn individual. He does not give the impression of having been able to suffer gladly those whom he considered fools. In terms of ensuring that the memory of Douglas was properly honoured it was his great patron, Sir William Hooker, who pushed Douglas' achievements into the spotlight and ensured that he would continue to be remembered. For posterity, Hooker wrote *A Brief Memoir of Mr David Douglas, with Extracts from his Letters* and published it in 1836. Personally, he was sure that Douglas' 'name and his virtues will long live in the recollection of his friends'.

Colourful annual reminders of his life still spring up all around us: in multicoloured *Penstemon*; vivid sky-blue *Nemophila* 'Baby Blue-eyes'; aptly named 'poached-egg flowers' (*Limnanthes*); curious *Mimulus*, or monkey flowers, growing everywhere, even in a ditch on the Isle of Lewis; and in the millions and millions of lupins. Stretching down twenty miles or so along the banks of the River Dee in Aberdeenshire are some of the original blue and pink lupins. The story goes that Queen Victoria's gardener was given a small sack of these strange and unknown seeds. He flung them in the river, and the seeds from this sack survive to this day, creeping further down the river as the years go by.

In Douglas' native Perthshire, the Royal Horticultural Society set up a fund to build a memorial to him. Over twenty-three feet tall, it is inscribed with a tribute to Douglas and a list of the various plants he introduced. The list of contributors to the monument is a testimony to the high regard in which Douglas was held. One of the saddest must have been Willie Beattie, head gardener at Scone and Douglas' old mentor. No matter that Douglas was on the cusp of being recognised as one of the most prolific and successful plant collectors of his era, it was less than twenty years since Beattie had been teaching him to weed, clip hedges, raise seedlings, squash caterpillars, nurture tender trees, and dig over the Perthshire land. Now his memorial would stand proud and lofty, its height matched by the tall trees he risked his life to find, which today surround Scone Palace.

Until his death, Douglas challenged himself to the limit of his endurance and evaporating physical strength. He admired in others the qualities he himself valued: determination, verve, drive and a respect for education.

CHAPTER 10

Thomas Drummond (1793–1835)

EARLY DAYS IN THE PACIFIC NORTH-WEST

———

Bright-eyed and full of bonhomie, Thomas Drummond seems to have shaken off misfortune as easily as a damp dog shakes itself dry. He gazes eagerly out from his portrait, looking for all the world like an exuberant young spaniel, with irrepressible bounce and a youthful conviction that he is immortal. He did indeed bounce, at the tender age of twenty-one, into the boots of the late impoverished George Don at Doohillock. As George Don junior poured venom on his recently deceased father's creditors, who, it appears, foreclosed their loans, Drummond, who was born in nearby Fotheringham at Inverarity, might have seen an opportunity. Alternatively, he could just have started work with Don a couple of years earlier and stayed on. Whatever the truth, there is no evidence that young Don and Drummond, who were only five years apart in age, ever communicated again after Drummond took over George Don senior's garden.

For the best part of ten years, Drummond tended the legendary garden, until he gave up and accompanied Sir John Franklin's expedition to find the North-west Passage. Who was rooting out likely personnel for Franklin's expedition? How did a gardener from Forfar come to appear on board as the botanist? It seems likely that it was the 'bush telegraph' of Sir William Hooker that alerted Drummond to the possibilities, and it was with Hooker, who was Professor of Botany at the University of Glasgow, that Drummond kept up a lifelong correspondence. Hooker was more than capable of pushing forward a likely lad to far climes; David Douglas was, after all, already in North America at his instigation.

John Franklin had already made his mark on exploration of the vast northern areas of Canada. Tales would have abounded about his first expedition from Hudson Bay to the northernmost point of the Coppermine River which had, on the return journey, suffered near complete disaster with two murders, a lawful killing in self-defence and cannibalism. With such experiences attaching to him it is amazing that anyone would want to accompany Franklin again. Perhaps just as amazing is the fact that Franklin wanted to undertake another expedition at all, especially as Franklin's wife was dying. She, however, implored him to go, tucking a silk hand-sewn Union Jack flag into his kitbag, which he was to drape over a pole upon sighting the Polar Sea. On 25 February 1825, the American paddle steamer *Colombia* set off from Liverpool with Thomas Drummond on board as an assistant botanist, having succeeded in getting a position on the expedition against strong competition. Dr John Richardson, who hailed from Dumfries and had trained in medicine at the University of Edinburgh, was on board as both naturalist and surgeon. Richardson was to accompany the party to the north once more, in search again of the North-west Passage, but Drummond was to set off to the south, to botanise, it being felt that the frozen Artic would elicit little in the way of flora.

Drummond accompanied the canoes and French Canadian porters of the Hudson's Bay Express route as far as Cumberland House, and after a short rest pressed ahead on 20 August, aiming for Edmonton House in Alberta, the next stopping-off point of the company. En route there Drummond established a pattern of collecting that was to continue for some time. During the day he would walk and stumble down the river bank, keeping within as close a distance behind the canoes as practicable, in order to be able to join up with the crew again for supper. He would pass most of the night huddled up in front of the fire cataloguing and preserving his plants (he was ever one for taking copious quantities of paper for pressing with him), and fall into one of the canoes at first light when they set off again, to curl up and sleep until breakfast time (which would probably be around mid morning), then rise up, eat, and start walking again. This routine he managed to maintain for the next 400 miles.

Arriving at Edmonton House, he left most of his luggage and tramped on north to Fort Assiniboine on the Athabasca River. With the weather closing in, and snow falling, they proceeded by canoe and on horseback to Jasper House, and then on yet another fifty miles upstream to the Rocky Mountain porterage, which entailed carrying the canoes overland to reach another river entry point. By now it was 18 October, and Drummond saw no point in going over the Rockies, especially as he was unable to carry most of his equipment. He hired a Native American hunter to look after and guide him and started trekking up the Snaring River. He reached a wintering post close to the Smoky River, at which point on New Year's Day 1826, at a deserted wintering station on the Baptiste River, his guide left him to it.

Drummond found himself quite literally up a creek without a paddle, or worse, without a shelter of any sort. No doubt tales of the previous Franklin expedition – the shudderingly cold conditions, with no game to speak of, and expedition members collapsing and dying from starvation and exposure – must have felt rather too close to home. He had a gun, but he had no books, and not even a pet animal for company. He wasn't quite sure where he was, and had no possibility of seeing any European with whom he could communicate well for at least four months. But he did have a happy-go-lucky boy scout disposition. He set about building a brushwood hut for shelter. He had no way of knowing that he was in for one of the coldest, longest winters for many years, but he settled down and appeared, unexpectedly for someone of happy and gregarious disposition, to have loved the sheer unpopulated wilderness around him. The grand welcome that had met Franklin and his accompanying expedition members at New York must have seemed no nearer than market day in Forfar. He later wrote, 'I remained alone the rest of the winter, except when my man occasionally visited me with meat; and I found the time hung very heavy as I had no books and nothing could be done in the way of collecting specimens of natural history. I took, however, a walk every day in the woods to give me some practise in the use of snow shoes. The winter was very severe and much snow fell in the month of March, when it averaged 6 feet in depth.'

Meantime, Franklin and a section of his men were holed up at what was now renamed by his loyal men Fort Franklin, in place of Franklin's intended name of Fort Reliance. A civilised routine had developed. Some men were detailed to tramp back and forth with correspondence to outposts on the Mackenzie River and the Slave Lake. The others spent the day either shooting and butchering or felling trees and splitting logs. In the long evenings, classes in reading, writing and arithmetic were commenced.

For Drummond, meantime, his longing for the first signs of spring seemed endless. When a very late thaw finally arrived, there was a major disappointment as his Native American guide refused to accompany him up into the mountains where he was hoping to get as far as a watershed of the Columbia River. Despite this setback he decided to set off alone, and was just about to start when a message caught up with him from Franklin telling him to return to York Factory on Hudson Bay by the following spring to meet up with the remaining members of the expedition for the trip home. This gave him less time than he had hoped; it was still a year away, but months would be taken up just returning to base.

April dawned and brought with it his guide. He donned the snowshoes in which he had spent the previous three months practising to walk, and set off for a rapid six-day trek to Jasper House, clutching his finds. He scooped up his tent, which had been sent up from Edmonton, and all the essentials a rugged plant collector needed: a bag full of paper for pressing, tea and sugar. At last, around Jasper House, he could pursue some serious collecting. His taxidermy skills had definitely improved as he sent back many carefully preserved birds and small mammals, but he was hesitant at applying these skills to larger animals, especially after a close encounter with a bear. Drummond's adventure is recounted both by William Hooker in his *Botanical Miscellany* and by William Gardiner in the *Flora of Forfarshire*.

The following sketch of an encounter with a grizzly bear, will give some idea of the danger of his undertaking, and furnish a sample of the graphic style in which he describes such an occurrence;

'Having crossed the Assiniboine River, the party halted for breakfast, and I went on before them for a few miles, to procure specimens of *Jungermannia No 17 in American mosses*, which I had previously observed in a small rivulet on our track. On this occasion I had a narrow escape from the jaws of a grizzly bear while passing through a small open glade, intent upon discovering the moss of which I was in search. I was surprised by hearing a sudden rush, and then a harsh growl, just behind me and on looking round, I beheld a large bear approaching me, and two young ones making off in contrary direction as fast as possible. My astonishment was great, for I had not calculated on seeing these animals so early in the season, and these were the first I had met with. She halted within two or three yards of me, growling and rearing herself on her hind feet, then certainly wheeled about and went off in the direction the young ones had taken, probably to ascertain whether they were safe. During this momentary absence, I drew my gun and the small shot with which I had been firing at ducks during the morning. I was well aware, [it?] would avail me nothing against so large and powerful a creature, but it was all I had. The bear, meanwhile, had advanced and retreated towards me three times, becoming apparently more furious than ever, halting at each interval within a shorter and shorter distance of me, always raising herself on her hind legs, and growling horrible defiance, and at length approaching to within the length of my gun from me. Now was my time to fire but judge of my alarm when I found that my gun would not go off as the damp had communicated to the powder. My only resource was to plant myself firm and stationary, in the hope of disabling the bear by a blow on her head with the butt end of my gun, if she should throw herself on me to seize me. She had gone and returned ten or a dozen times, her rage increasing with her confidence, and I at any moment expected to find myself in her grip, when the dogs belonging to the brigade made their appearance. But on beholding the bear, they ran back with all possible speed. The horsemen were just behind, but such was the surprise and alarm of the whole party, that though there were several hunters, and at least half a dozen guns among them, the bear made her escape unhurt, passing one of the horsemen whose gun like mine missed fire and apparently intimidated by the number of the party.'

Drummond had learned his lesson, he would make sure he looked after his gun more carefully in future, and he returned to collecting his moss. The best method to get the better of bears, he noted with panache several months later when writing up his journal, was to rattle one of the tins he carried specifically for seeds and botanical specimens. No further reference was made to encounters with bears.

On 16 June his luck changed. He found an ageing Canadian, who was detailed to take the Hudson's Bay horses up to the summer pastures, and he quickly tagged on behind, at long last poring over grasslands which were richly carpeted with flowers – although annoyingly they were equally buzzing with mosquitoes. Finally, Drummond had crossed over the Rockies, and was able to drink a large draught of water from a tributary of the great Columbia River, pleased to know that the water might have descended from the northern ice-bound areas and was passing on its way to the Pacific. In the meantime, Drummond and Douglas were about to meet in person, although they had each already heard about the other during the course of their travels.

In March 1827, while David Douglas was leaving Fort Vancouver for the long journey over the Rockies and to catch the boat at

York Factory for home, Drummond left Edmonton House on the track to Carlton House. As the local Native Americans were in no mood to be hospitable to the Europeans, Drummond's guides led him with their two dog-sleighs into unfamiliar territory to avoid conflict. It was hardly an auspicious start. Lost for much of the time, their visibility blinded by storms, they could hardly aim accurately at the scarce game they glimpsed. A skunk provided one meal, a deer skin, carefully preserved by Drummond for his collection, was scraped clean of flesh, and when that wasn't enough to avert their increasing need for calories to offset the sapping effects of the extreme cold, they ate the skin as well. The dogs were too exhausted to pull the sleighs, so Drummond and his guides stood in as haulers. One very long month later, on 5 April, the exhausted party reached the safety and comfort of Carlton House.

A month after that episode, which Drummond had spent cataloguing his collections, on 5 June, he encountered David Douglas who had also followed a detour to avoid aggressive natives. From now on they were in each other's company. For their final few months in Canada, both he and Douglas were swept up in the rapid journeying of the men of the Hudson's Bay Express who had little truck with meandering plant collectors. Their jobs and reputations were founded on covering the miles in as short a time as possible.

By the second week of September, Douglas and Drummond were aboard a Hudson's Bay ship, the *Prince of Wales*, ready to embark for England – not for Drummond the hero's welcome awaiting Franklin and Richardson in New York. Douglas and Drummond were accorded merely a couple of hammocks on board the regular Hudson's Bay boats which plied between Canada and Britain. Franklin and Richardson had in the meantime met up from their separate explorations on the extreme north coast and travelled together back to New York for a week of receptions. They were presented with such delights as a splendidly bound volume of *Cadwallader Colden's Memoir at the Celebration of the Completion of the New York Canals* and a medal just struck to commemorate the opening of the Erie Canal. On 1 September they embarked on the paddle steamer back to Liverpool, reaching England twenty-six days later. Drummond and Douglas, however, arrived back by the slower Hudson's Bay ship at the beginning of October. Douglas experienced brief celebrity fame, which he found difficult to enjoy. Drummond, being a more ebullient character, might plausibly have enjoyed public acclaim. Instead he was honoured through the pages of William Hooker's *Flora Boreali-Americana*, a tome which was based largely on the plants brought back by himself and Dr John Richardson.

At the end of 1827 Drummond was at a loose end, most probably tending his neglected garden in Forfar, which his wife had managed to maintain in his absence. However, both the maintenance of the garden and his stay in Scotland were short-lived. when he seized on the opportunity to involve himself in the newly planned Botanic Garden in Belfast. He arrived there a year before the garden ground was purchased, and stayed for two years in the employ of a committee who were in charge. It was an unhappy period all round and there was underlying tension from day one. Drummond blamed the committee of the Belfast Botanic Garden for lack of funds, and for employing cheap labourers instead of apprentice gardeners. On the other hand, some of the good burghers of Belfast resented the fact that Drummond was a Scot. That no native Irishman was up to the task weighed unhappily on their shoulders.

One of Drummond's supporters was Mrs Katherine Templeton, the widow of a well-respected Irish botanist. She spoke out on Drummond's behalf, and took a personal interest in Mrs Drummond and their children, who must have been still under ten years old, although no records appear about his marriage or children. Mrs Templeton seems to have been a worthy and intelligent woman, who was not averse to some straight talking. It was she who implored Drummond to refrain from indulging in strong drink – another bone of contention between himself and the other members of the committee – and she was happy to write to the ever-interested William Hooker that Drummond had abstained from 'strong liquor' as promised.

Drummond, in the meantime, had made plans. He would leave Belfast, and if Hooker could put his name behind a proposed plant-finding expedition to be carried out by Drummond in the southern states of North America, he would consider going. Drummond cannot have fallen foul of everyone in Belfast as one of the committee members, the treasurer Valentine Whitla, pledged a personal £20. Mrs Templeton defended Drummond's work and described the garden which he had created in Belfast as 'very pretty'. Drummond was sorry to leave, but unrepentant. He had railed against the restrictions of being starved of both cash and trained assistants, and warned that a botanic garden could not be created without either. He was proved correct. The next curator lasted an even shorter time, and the garden withered. Gradually, it metamorphosed into a public park.

Initially, Drummond turned his ambitions to the west, aiming for Santa Fe, New Mexico. Less optimistic characters might have fretted about leaving behind a wife and children. Drummond was indeed concerned for their welfare and, of the money he raised to sponsor a trip to America, he took only what he absolutely needed, ensuring that enough was left over to keep his family from the breadline. The long-suffering Mrs Drummond seems to have been the silent pillar of strength here, and it was she who kept the family together, moving back close to her home area in Angus. This time, however, she was not returning to the Doohillock garden. She was now just responsible for her children, and not dealing with the garden assistants as she had apparently done when Thomas had been trekking over North America.

Drummond gathered together sponsorships from the Botanic Gardens at Edinburgh and Glasgow, as well as from many private subscribers, to collect plants for sale at the rate of around £2 per hundred, to be distributed by William Hooker. He arrived in New York on 25 April 1831, to a quieter reception than before, when he was dazzled by John Franklin. He was clutching his letters of introduction. These were entry visas into the world of influential men in North America and Drummond had secured some of the best, covering most known areas of the south and west of the United States.

He had letters to 'Mr Astor, the Head of the United States Fur Company, whose influence extends from the Mississippi to the Pacific Ocean', and for good measure, his rival, 'Nicolas Garry, Deputy Governor of the Hudson's Bay Company, a passport which would command every thing necessary for the furtherance of Mr Drummond's object, in case he should find it desirable to enter their extensive possessions from any portion of the United States.' He had letters of introduction to Messrs Manning & Co. of Mexico, Messrs Baring of Baring Brothers & Co., and lists of local names from Glasgow. Many Scots had contributed their influence: his associate Dr John Richardson from the Franklin expedition; the British Museum; and Charles Lyell, a landowner and noted naturalist from nearby Kinnordy close to Drummond's birthplace, to name just a few. Lyell was to remain supportive of Drummond's family for years to come.

One of these letters had been carefully scripted by William Hooker, therefore ensuring a meeting between Drummond and Dr John Torrey, an American botanist who became professor of chemistry and natural philosophy at Princeton University, a couple of years his junior at thirty-five years of age. Torrey had already studied plants from an expedition to the Rockies as well as many from an expedition down the Missouri River on a steamboat in 1819. So he already had a wide working knowledge of the plants in those areas, having catalogued them from 1824 to 1828. Torrey reported that Drummond had already ordered 'two tons of paper to arrive ready for him at New Orleans'. He also put out the word to his acquaintances that Drummond would be happy to add any of them to his list – Drummond needed badly to find plenty of potential buyers for the many plants and seeds he had every intention of collecting. For good measure Drummond presented Torrey, on behalf of both himself and Hooker, with a selection of the mosses he had collected on the North American trip with Franklin, carefully catalogued by Hooker.

All in all, this was a more favourable commencement to his journey than that of David Douglas eight years before. On that occasion, Douglas had been instructed by Hooker to present Torrey with seeds from Britain. Torrey not only showed great irritation with Douglas, who appeared to be vague as to the contents of the parcel, but fired off a furious letter to Hooker complaining that Douglas could not provide satisfactory answers regarding the seed package and 'he is such a liar I know not whether to believe him or not'. But Torrey reckoned that on the evidence of the quantities of paper alone, this newly arrived Scot meant business.

He most certainly did. His intention was to take the trail which had been established by local Native Americans, followed by Spanish explorers, and, exactly ten years earlier in 1821, had evolved into the legendary Santa Fe Trail. A certain Captain William Becknell had blazed a trail from Missouri, intending to reach Mexico. But arriving in Santa Fe, he found himself surrounded by locals who bought just about everything he was carrying. The profits from such rapid sales sent him scurrying back for replenishments, and others quickly followed his example. But it was prairie plants which attracted Drummond, reading as he might well have done from Becknell's diary that 'A continual and almost uninterrupted scene of prairie meets the view as we advance'. Although Drummond already carried letters which would guarantee him a protected passage through the zones which were under the control of the two great fur trading companies, who often regarded themselves as unofficial feudal chiefs, he also visited the British Consulate in New York, who gave him an introduction to the British 'minister' or consul at Washington. Outposts for military protection were being rapidly built along the Santa Fe Trail to protect travellers and wagon trains, so credentials were useful.

Thomas Drummond set off southwards. By this time, travel between the major cities of the east coast was easier, undertaken either by wagon or, more easily, by boat down the many rivers.

He set off for Philadelphia, buying the maps that he required for his intended destination, Santa Fe. He also sent some seeds back to Hooker from the area, then proceeded on to Baltimore, Washington and Fredericksburg. He then set off, minus almost all his luggage which was taken by wagon in order for him to travel as lightly as possible while traversing the Alleghenies, the heights of which he dismissed as being mere hillocks in comparison with the Rockies. Intent on gathering plants, it was the first of a list of disappointments: the specimens rotted, being too full of sap to preserve so early in the season. He was also disappointed to find that the wagon travelled 25 miles per day, and as he was required to catch up with it every night in order to catalogue and press his specimens, he found he had not enough spare time for collecting. But he was still more or less on track with his plans, fairly astonishing as these had been laid largely in Angus and Glasgow, born of dreams in Belfast.

He arrived at Wheeling, West Virginia, on the Ohio River seventeen days later, and waited a week for more of his luggage to arrive, which probably included copious quantities of paper. He had intended to meander down the Ohio on a small boat, being able to stop at will to explore the local countryside and generally take his time about it, systematically quartering the areas for maximum coverage and an efficient search. But small boats had been superseded by large steamboats and nobody had a boat available to hire. So he boarded the steamboat to Louisville. Lying on the boundary between Indiana and Kentucky on the Ohio River, the town sweltered during those months of June and early July, and was riddled with disease. This was exacerbated by hordes of travellers moving west in search of land, carrying with them virulent infections. Within three days he was poleaxed by fever and laid up for ten days. Feeling marginally better he staggered back on to the steamboat and in four days reached St Louis on the border between Missouri and Illinois, where he immediately succumbed again. He sought medical advice, which helped little. He lost many specimens. Unable to walk for two weeks, he could not search for nor bring back the many flowers which scattered the countryside and which he was desperate to collect and dry. He got up briefly but then was reduced to lying in bed with severe diarrhoea, and in a letter to William Hooker, he described his state as 'little else but skin and bone'. Finally he managed to find enough strength to make short excursions although he had to rest six times in the space of about a mile. Desperate to make up for lost time, desperate to find seeds in order to satisfy his sponsors, and desperate to press on to the west, he pushed himself to his physical limits. The paper that he had so carefully organised to be sent by Messrs Dennistoun of Glasgow to New Orleans had to be forwarded up to him as the paper he had posted to St Louis had not arrived. He had to buy paper at an extraordinarily high price and the firm pasteboard which he required for stability was unobtainable.

But all this paled into insignificance when he realised that he was too late to proceed on the wagon trail with the fur traders as they had left during the first week in May. The area he wished to get to was the same area as the fur traders' hunting grounds which were at the source of the Missouri River. He resigned himself to making the most of his enforced stay in the area, and settled down to rest and explore his surroundings. By December, his health partially restored, he was packing up gigantic grasses, which he described curiously – since New Zealand had been little botanised at this stage – as having the habit of New Zealand flax. He boxed up hundreds of living plants and penned a long tale of woe about the frustration of not yet being in prairie country, about the stunted and uninteresting trees adjacent to St Louis, and most of all about his ire concerning the unhealthy climate. 'Fever is universal about here, not one of 50 escaping either among natives or strangers,' he complained to William Hooker.

With very few travelling options open until the following year, he proceeded down the Mississippi to New Orleans by steamboat, which again was frustratingly fast. From New Orleans on 3 January 1832, he sent dozens of seeds and specimens, collections of oak acorns and pine cones, reptiles preserved in spirits, mosses, shells and insects. During that spring he ventured out in the neighbourhood of New Orleans and sailed to the opposite shore of Lake Pontchartrain, finding copious quantities of shells, insects and *Cryptogamiae*, a type of flowerless plant.

While in New Orleans he had sighted plants collected in Texas in 1828 by Jean Berlandier for a Swiss collector in Geneva, de Candolle. He had never seen anything similar and was now determined to go there, wending his way up the Brazos River, the mouth of which is west from present-day Galveston. The most southerly of the states of North America had been a long-running argumentative hothouse between, variously, the Spanish, the French and the Mexicans, the last-named of whom languished in the dying days of control by Spain, a fading colonial power. Texas, in the days when Drummond pushed inland, was still an unstable frontier. Settlers had arrived from Mexico and the south because of Mexican government incentives in the form of sheep and lambs. By the time that Drummond reached the area, pressing in from the east, there were more than 30,000 settlers, but in such a colossal area they were no more than pinpricks on the land. Drummond arrived in the early stages of skirmishes for Texan independence.

For Drummond, however, it was the Native Americans, the weather and cholera which delivered the worst blows. As for hostile natives, he missed one possible inland expedition, which proved to be a blessing in disguise as everyone was massacred, though Drummond appears to have been oblivious to the near escape. In terms of weather, as well, Drummond couldn't have chosen a worse time. The spring of 1833 was known as the year of the 'great overflow'; floods covered the entire area and he reported that the river was so high it was a week before they could even go twenty miles upriver. As the floods receded he was left in the small hamlet of Velasco.

As he collected round the area of Velasco, at the mouth of the River Brazos, he complained bitterly of the light-fingered habits of the local settlers with whom he rubbed shoulders. He moaned of skinning birds with a common old penknife, 'not worth two cents, and that even this shabby article I could not have kept had the natives seen anything to covet in it. I am obliged to leave behind my blanket and a few clothes that I have bought, because of the difficulty of carrying them, though I feel pretty sure that I shall never see them again. These trifles I only mention to give you some idea of my present situation, they do not affect me much, except as prevented me from pursuing the objects of my journey with the success that I could wish.'

The most deadly danger was invisible. Cholera killed almost all of its victims, with the notable exception of Drummond. While still in Velasco, Drummond caught the disease, but a prompt self-administered dose of opium appeared to save him, or so he deduced. However, his recovery period in the immediate aftermath was almost more of a problem, owing to lack of nursing or care. He was so weakened by illness that he was quite unable to venture forth and search for food, and little assistance was around. With no more than a handful of survivors, all too weak to help each other, he nearly starved to death. 'The Captain and his sister died, and seven other persons, a large number of the small place, where there are only four houses.'

Once again, however, Drummond rallied, boosted by the great number of flowers and new specimens he was finding and the germ of an idea that was forming. Regardless of cholera, floods, thieves, war and hostile natives, he saw Texas as a land of opportunity. He had been infected not only by illness, but by the 'American Dream'. He would send for his family, or go back and collect them. He would settle in Texas. A man with so much bounce couldn't be downcast – not by three near-fatal illnesses in less than a couple of years, and certainly not by the facts. He disregarded so much of what was clearly obvious. No matter that it was a time of great lawlessness in the area, with most of the settlements being classic one-horse towns where arguments were settled by fist and bullet. San Felipe de Austin in Texas was, in 1832, a settlement of about thirty families, with several stores and two taverns where travellers were offered very simple fare.

At the end of October he wrote from San Felipe de Austin informing Hooker that he had forwarded specimens of fruits, plants and curious lizards, and announcing that his journey so far in that area had produced about 150 species of plants, therefore totalling a haul in Texas of nearly 500. He decided to stay in the Galveston area over the winter, and was beginning to talk about returning to Scotland in a year's time. In the meantime he was determined to collect seabirds on Galveston Bay. He spent the entire month of January on Galveston Island which he said was reputed to be one of the greatest habitats for seabirds on the whole coast. It rained incessantly. The island was uninhabited. He found almost nothing to eat but he did manage to return with 180 specimens, 50 of which he had never seen before. The mainland area was throbbing with Native Americans who were once again posing a real threat, but it did not stop him finding an old canoe and paddling from Galveston Bay upriver, covering between eighty and one hundred miles all alone with hardly any provisions. If he thought he could eat off the land, he was mistaken. Food was almost non-existent, and he was virtually starving. No settler had a surplus as the previous year's crops had failed, owing to the floods, and famine was a real threat. But, he wrote home, his health was restored and excellent. This optimism was short-lived.

By the end of September he had abandoned plans for crossing the mountains into New Mexico and reaching Santa Fe, as the intervening country was full of hostile natives. While still in San Felipe de Austin, he joined a wagon bound for Gonzales in Guadalupe. On 20 December, he was back in New Orleans, and wrote reassuring Hooker that his collection of plants from Texas now amounted to 700 species.

On Christmas Day, Drummond acknowledged a kindly letter from Hooker which encouraged him to go to Santa Fe directly. Hooker would have helped sponsor Drummond's voyage on the understanding that he would search out plants from the Santa Fe area, and was reiterating this message, almost as if Drummond had carelessly missed a wagon connection. It is difficult to tell if

Hooker had already received the news about the difficulties of travel. Hooker was an extremely humane man, and it seems unlikely that he would have forced someone for whom he had so much respect to face danger deliberately when he knew the outcome could so easily be death. Drummond's demise would, even to a cold-hearted man, have seemed very much counterproductive to plant collecting. Whatever the truth of the matter, Drummond stood firm: he was quite determined to return to Scotland. Once more, illness had diminished his physical strength, and even his resolve showed signs of cracking. This time he reported a violent attack of diarrhoea, accompanied by ulcers, describing himself 'as almost like Job, smitten with boils from head to foot, have been unable to lie down for several nights'. He also had a badly infected thumb which he was in danger of losing.

Still, it was an agonising decision. On the one hand he was full of plans for the future. He would buy a league of land for $150 and purchase a dozen cows and calves at $10 each, with every possibility that it would make him more independent than he could ever hope to be in Britain. From that healthy home and financial basis he would be able to explore from Texas to Mexico City and west to the Pacific, an area which he reckoned would occupy him for at least seven years. On the other hand, he was aware he had not covered the area which he had promised Hooker he would explore, and indeed had been his intention.

His new plan was to leave on 1 January. He managed to obtain a passage on the ships sailing for Apalachicola, Florida. Arriving there he wrote again to Hooker on 9 February 1835. He had not found a way of going to the south of Florida, so had decided to sail, taking a ship bound for Key West and eventually England. Waiting in Glasgow, Hooker received three boxes in June 1835, containing not seeds but meagre belongings with a death certificate, that of Thomas Drummond. It was dated 11 March 1835. Hooker gave no indication of the reason for Drummond's death, nor whether any reason was stated. By June, though, Hooker had noted that number 3,441 *Phlox* 'seeds sent over the early part of the year 1835 soon vegetated; the plants blossomed most copiously and with equal profusion and brilliancy of colour, whether in the greenhouse or in the open border, and it bids fair to be a great ornament to the gardens of our country. Hence and as it is an undescribed species, I am desirous that it should bear the name and serve as a frequent memento of its unfortunate discoverer.' He named it *Phlox drummondii*.

Back in Angus, Mrs Drummond received the news she had been dreading. She worried for her children, but Thomas' career had been followed and admired greatly by many. Just as Brodie of Brodie Castle among others had gathered together to help out the widowed Mrs Don and her children, so Mrs Drummond benefited from the reputation of her late husband through the generosity shown to their son. Charles Lyell, the Forfarshire landowner and naturalist who had been so interested in Drummond's work, stepped forward and paid for the education of Drummond's son at the Dundee High School. *Phlox drummondii*, along with many of the other great flowers of the plains explored by Drummond, such as *Coreopsis*, *Rudbeckia*, *Gaillardia* and *Penstemon*, still burgeon out of window boxes across Europe.

CHAPTER II

John Jeffrey (1826–1854)

TREES BUT NO TREASURES IN GOLD-RUSH COUNTRY

If the head gardener at the Edinburgh Botanic Garden had been impressed by the young Jeffrey when he volunteered to climb a high tree to cut down a dangling, damaged branch, his opinion was confirmed when Jeffrey won outright the prize offered to gardeners for the best collection of dried plants in the Edinburgh area. Other than this, little enough is known about his tenure at the Botanic Garden. To this day, John Jeffrey remains a shadowy character in the annals of Scottish plant collectors. As Coville put it in his paper for the Biological Society of Washington in 1897: 'Among the most obscure of the scientific explorers of the Pacific North West is John Jeffrey, whose journey is sometimes referred to as the Oregon botanical expedition. Even though he made valuable contributions to the botany of the Pacific North West and his name is associated with several plants, only one American writer at the turn of the century attempted to trace the course of events surrounding this little-known English botanist and explorer.'

It is hardly surprising that few tried to trace him: Jeffrey in his diaries left little in the way of personal detail, and indeed ceased writing altogether. Details about his life are therefore sketchy, unlike Menzies, Douglas or Drummond who took pains to write copiously and offered plenty in the way of opinions on their fellow travellers, or indeed the organisations by whom they were paid. Jeffrey was a lad from a backwater of Perthshire, quiet, unassuming and unassertive. He spent his childhood and early life but a stone's throw from one of the prettiest lochs in a romantic and relatively prosperous area in the parish of Clunie. The island in the centre of the loch allegedly housed the Admirable Crichton, a poet, scholar and traveller whose short life ended in mysterious circumstances – although brawling and murder are hot favourites – while in the service of the Duke of Mantua in Italy in 1582.

John Jeffrey was born on 14 November 1826, the eldest son of John, an agricultural labourer, and Helen Ambrose, three years his junior. By 1841 they had three more children: James, three years younger than John, Janet three years younger than James, and Elizabeth three years younger again. By now, father Jeffrey and his family had moved to the south of Fife, at Lochore, at least 50 miles away. Later he returned to Clunie in Perthshire, until finally moving permanently to Fife. John's two sisters, Janet, the family housekeeper, and Elizabeth, a 'scholar' or schoolgirl, stayed with their father. They remained firmly within their social stratum.

At the age of fourteen, future explorer John Jeffrey was no more than an agricultural labourer, or male servant according to contemporary records, almost certainly living in the rough bothy on the farm. This household consisted of David Cochrane and his wife Elizabeth who farmed Wester Tullyneddie, at Clunie. The Cochranes employed two female farm servants, both aged twenty, another labourer, Donald Stewart aged twenty, John Jeffrey, and finally James Miller aged ten.

This area remains one of outstanding beauty and the cluster of lochs along the road from the ancient cathedral town of Dunkeld to Blairgowrie is famous. The *Statistical Account* of 1841 states:

> Wet grounds have been drained, rough grounds cleared, stone fences built, and hedges planted. Lime is brought from the quarry and to lime kilns upon the Gourdie estate, belonging to David Kinloch. Good permanent soil has formed where was no soil

before; green crops begin to be raised, and a regular rotation of crops begins in some places to be understood. All was due to lime fertilising the soil.

With one or two exceptions, the farms in general are small. Few of them probably rent above pounds 200 Stirling. The staple grains are oats and barley.

In fact, more grain was produced than was necessary to meet the local contribution to the minister's stipend, a good sign. Another encouraging omen for the area was that the teacher at the local school had a class of forty youngsters (boys) who were taught reading, writing and arithmetic, with two learning Latin. The teacher received the maximum salary. For some girls there was a school in the parish, taught by two females, where the fifty-one pupils learned knitting and needlework. A library had opened two years earlier, paid for by local subscription. The poor benefited from the interest on a legacy left by the late minister, the Reverend William MacRitchie. This allowed coal to be bought for 'the industrious poor [not beggars but the genuinely destitute] of his native parish'. The local farmers drove the coal free of expense.

From this well-ordered and fairly prosperous pocket of Scotland, John Jeffrey seized a series of opportunities. Somehow he made the transition from farm labourer, to gardener at the Royal Botanic Garden in Edinburgh where he worked for about one year before becoming a hand-picked plant collector destined for the far west of the United States of America. It was all accomplished within ten years, a truly meteoric career rise, but there are no clear clues as to how he transferred from life as a lowly farm hand even to a gardener.

There is, however, circumstantial evidence that might be relevant. There were two local landowners who were both keen agricultural improvers: James Speid, a writer and ex-Provost of Brechin, who had inherited Forneth estate from his brother-in-law John Binny, 'sometime merchant in Madras'; and David Kinloch of Gourdie House to the south. Both estates sat in such good positions that the local minister of the time was driven to say that they were both 'beautifully situated' and that the prospect from Gourdie 'is delightful'. Both lairds had cleared hundreds of acres of heather, broom and brushwood and the land was 'now covered with beautiful thriving plantations of fir and larch'.

Close by also lived Patrick Matthew, born near Scone in 1790, described by his biographer, W.J. Dempster, as a nineteenth-century gentleman-farmer, naturalist and writer. Charles Darwin acknowledged that Matthew's deductions on laws of natural selection had pre-empted his conclusions, going as far as to say, 'I freely acknowledge that Mr Matthew has anticipated by many years the explanation which I have offered of the origin of species under the name of natural selection'. In 1831, Matthew had published a book on *Naval Timber and Arboriculture* with an appendix on natural selection.

Thus Jeffrey had spent his most formative years surrounded by 'new' trees. He would have heard much chat locally about the pros and cons of such new plantations. He would also have heard much about the wealth tied up in such plantations, the high financial expectations placed on such rapidly growing trees. Rumour would have reached him about the exploits of David Douglas and his reports of gigantic trees, all of which suited Scottish hillsides and climate. Would such stories have stirred up ambition in his soul? Would he have thought that if Douglas could make the leap from lowly gardener to explorer, he, Jeffrey, could do the same? Might he have felt that the world was his oyster?

Such exploits and discoveries had also fired the imaginations of landowners who saw not only how to make a fast buck from trees which would produce timber in record time, but also the attraction of specimen trees which would enhance their estates. It was George Patton, who later became Lord Glenalmond, of The Cairnies – a house now incorporated into Glenalmond College – fifteen miles from Perth, who initiated the idea of sending out a hand-picked collector from Scotland to bring back plants just for Scottish landowners. With that germ of an idea, he cast around for others to back the scheme. He spoke to Professor J.H. Balfour, Director of the Royal Botanic Garden in Edinburgh, who thought it a splendid idea and, on 22 November 1849, a meeting was held to suggest the names of willing subscribers. Events then moved rapidly, another meeting was called within the month, with the formation of a Grand Committee on 30 January, and the first official meeting of the Oregon Association was held on 6 February.

With some hiccups and copious advertising in such journals as the *North British Agriculturalist, Gardener's Chronicle, Edinburgh Evening Courant, Scotsman, Edinburgh Advertiser, Glasgow Courier, Dumfries Herald* and *Bell's Life in London and Sporting Chronicle*, donations began to pour in. Encouraging Queen Victoria's consort, Prince Albert, to kick-start the proceedings by donating £10 would be a considerable coup. He was duly placed at the top of the list of subscribers' names, which then rapidly became a roll call of the aristocratic. In careful deference to the status of such titles, and in strict hierarchal order, the list of subscribers encompassed the Dukes of Devonshire, Montrose, Sutherland, and Roxburghe, followed by Marquises and Earls of Aberdeen, Burlington and Hopetoun, Lords Kinnaird, Lovat and Walsingham, and then the various admirals, judges and societies which also contributed.

The solid, ordinary Scottish names in the Association's list of subscribers, such as Barclay, Anderson and Glendinning, were enlivened by the addition of the Vilmoran brothers who were grandly inscribed as *pépiniéristes* (nurserymen) from Paris. The well-established firm of James Veitch & Son of Exeter, the Royal Caledonian Horticultural Society, Professor Lindley from the Horticultural Society of London and Messrs Dickson and Co., well-established nurserymen in Scotland whose business offered training to many a budding plant collector, all sent in a subscription.

Sir Robert Menzies, a descendant of the family which had employed Archibald Menzies' father, at the ancestral castle, Menzies Castle, by Aberfeldy in Perthshire, contributed, as well as Sir Thomas Gladstone, who registered under his Scottish home, Fasque, Fettercairn, Kincardineshire.

The generosity of subscriptions varied widely. Nurserymen who

clearly had a stake in importing new plants appeared to think it
was well worth their while to stump up, and paid more than most,
but outstripping these nurserymen, and the vast majority of private
subscribers, was the Duke of Buccleuch of Dalkeith Palace, who
subscribed a colossal £60. To put this in proportion there were
281 subscribers pledging the total of £1,440, with most putting
their hands in their pockets to the tune of £5 or, at most, £10.
Most just wished to get some of the seeds of the fabulous bounty
that they hoped was coming, and, of these, most appeared to be
interested in trees. On 20 February, the committee advertised for
a suitable collector, and petitioned the Admiralty requesting a
berth on a suitable ship going in the direction of the west coast
of North America. In the meantime, there were several applicants
for the post. By 28 February 1850 the committee, however, decided
to appoint 23-year-old John Jeffrey who had strong support in
the form of Mr James McNab, Principal Gardener at the Royal
Botanic Garden in Edinburgh. Additionally, Professor Balfour
had earlier hinted strongly that 'he had already someone in mind
for the post'. No doubt the two men had more or less stitched up
the appointment in tandem, and this would have made eminent
sense, as they were well acquainted with Jeffrey and he was readily
available and on site. He was tried and trusted.

Even a committee as diverse as the newly formed one must
have found it a difficult post for which to make an appointment.
They were sending Jeffrey to the same area covered by David
Douglas, and Thomas Drummond had been known to venture
there too. Both, it was only too well known, had died. Few of
the committee would have travelled further than London, most
had almost certainly never crossed the Atlantic and would have had
almost no idea of the conditions of travel in the remote northern
territories of Canada. In the meantime Sir John MacPherson Grant
of Ballindalloch, a committee member, approached one of the
great old warriors of the Hudson's Bay Company, Edward Ellice,
who responded promptly on the company's behalf by offering to
take a collector on one of their ships to York Factory and then

escort him across the continent. Even more generously, as was carefully recorded within the books in Edinburgh, the Hudson's Bay Company had also offered to supply all his needs while in their territory and supply him with money if he went outwith their limits, up to the amount of credit the Association would grant him.

John Jeffrey was ordered to collect a range of conifer seeds, any other plants of interest and commercial value, and a copious amount of beetles. He was to keep an eye open for any subjects of natural history, shoot the unfortunate bird or animal, and skin and dry it before dispatching it to Edinburgh. He was also to keep a diary.

He duly signed the contract for his three years of exploration on 29 May 1850, just six months since a tentative first meeting had been arranged between George Patton of Glenalmond and Professor Balfour to sound out the idea of sending their own collector.

Almost immediately, Jeffrey set sail for London and it appears likely that he had been instructed to call upon Sir William Hooker and Dr Lindley of the Horticultural Society among others, no doubt to glean from them any helpful hints about gathering and packing of seed, or indeed what types of plants and seeds he might make a priority. Evidence of these meetings was in the letters of thanks which flowed in his wake from the committee secretary in Edinburgh. No doubt the Edinburgh committee were anxious that Jeffrey should be as well prepared as possible, although for the young Scot it must have been a daunting experience to present himself before such eminent persons, when he had no doubt spent most of his working life out of doors and dressed in a gardener's smock. Jeffrey, for his part, would have been only sketchily acquainted with the journeys of Archibald Menzies and David Douglas. However, the journeys of both men would have been well known to William Hooker. Neither Menzies' nor Douglas' accounts had by this time been published, and the information released to Jeffrey by Hooker would most likely have been carefully filtered, delivering only the bare bones of the dangers of the journey ahead, but paying singular attention to the types of seeds required.

For the kindly, patriarchal figure of Sir William Hooker, there must have been mixed images and thoughts as he viewed the keen young Jeffrey, so similar in age, stature and accent to Douglas. He must have wondered if he would ever see him alive again.

On 6 June 1850, Jeffrey climbed up the gangplank of the *Prince of Wales* which was moored at the East India Dock, and sailed for Canada. His last sight of Scotland would have been the harbour at Stromness, on Orkney, from where the Hudson's Bay Company typically acquired the tough men for their outposts right across the Northern Territories. On 12 August, they landed at Five Fathom Hole, off York Factory, and Jeffrey left, two weeks behind schedule, under the wing of chief factor John Lee Lewes.

The Hudson's Bay Company was famous for being one of the most successful companies in the world at that time, for their Scottish nepotism, as a post was always found for a likely lad related to an employee, and for their sheer tightfistedness in conducting their business. Although the company, specialising in beaver skins for the lucrative European market, resented any European intrusion into the millions of acres that they regarded as their own fiefdom, a youngster who was to be strictly under their control posed no threat. And so Ellice would have been more than happy, in the spirit of the company's skill at extracting cash from any likely source, to relieve the fledgling 'Oregon Botanical Association' of £18 10s for Jeffrey's passage, and for good measure a further £10 10s for 'messing' (food).

Jeffrey had been allowed just £20 to purchase books and instruments; his outfit, consisting of strong coats, breeches, stockings and boots, cost £49 10s 2d; his wages for the three years totalled £267 and his life insurance premiums cost the best part of £100. With the abysmal record of life expectancy of previous explorers and seamen in this region, it was a wonder anyone would insure him at all.

A week following his twenty-fourth birthday, on 20 November, Jeffrey arrived in Canada. As the enormity of the travelling expected of him began to sink in, and the bone-chilling air of the area infiltrated his new clothes, he might well have had a suspicion that his financial rewards were not quite as generous as he might have thought less than a year previously, dreaming of adventures in the gentle, steamy comfort of the potting sheds at the Edinburgh Botanic Garden. Even if it had been possible for the Oregon Association to communicate with him rapidly, he would still without doubt have been kept fully in ignorance of the dizzying totals of the subscriptions, at that point already £950. That colossal figure would itself rise, on account of further subscriptions over the next five years, to £1,667 8s 2d.

Jeffrey had already taken part in 'expeditions' from the Royal Botanic Garden in Edinburgh with Professor Balfour, but these had consisted simply of robust walks over Highland hills and glens. Perhaps he imagined that these had been in some way a sound preparation for this type of life. They had roughed it to the extent that travel to the closest point from which to explore a wilderness was by third- or fourth-class train. Most likely, the first time that Jeffrey had ever set foot on a boat of any size would have been his trip down to London from Leith docks, near Edinburgh. This new experience must been a huge shock to his system. However, the Hudson's Bay Company men were well used to young lads, fairly wet behind the ears, arriving and being thrust into the harsh environment, and it appears that Jeffrey was well received. His physical fitness was put immediately to the test, as they trudged and canoed to Norway House near the north-east end of Lake Winnipeg, arriving three weeks later. With one day's rest, they took off the next day for Cumberland House on the Saskatchewan River which was the destination of John Lee Lewes. Here, Jeffrey fully realised that the sheer cold and depth of snow were beyond anything he had previously encountered. He was left to kick his heels until 3 January 1851, waiting for the Express, the Hudson's Bay Company's cross-country canoe and porterage service, with whom he was to travel, following the well-known route right across Canada. He was travelling virtually in the same footsteps as David Douglas nearly twenty-five years before.

They made for Edmonton House, to be joined by chief trader Robert Clouston, who had also been waiting there for some time. The reasons for the Express's delay now emerged and must have given Jeffrey even more cause for nervousness. Not only was the snow exceptionally deep on the mountains, therefore delaying the beginning of the trip, but also Clouston had wind of the hostile intentions of the Blackfoot tribe, who had staked what they thought an entirely justifiable claim overlooking the mountain passes through which the Express had to travel. Jeffrey and Clouston left Edmonton House in early March, reached Jasper House on the Athabasca River two weeks later without mishap, and spent a relieved couple of weeks there with Colin Fraser, then in charge of the post.

On 7 April 1851, heading up his letter Jasper House, Rocky Mountains, Jeffrey began his report on his travels so far. He explained that things had not gone quite according to plan. He had understood that every ship arriving from Britain was met by a brigade of voyageurs who then proceeded over to the general area of Fort Vancouver, rather like a regular travel service. This was not so, as the brigades indeed met the ships, but then offloaded their furs, loaded up with supplies and returned to wherever they were based, so scattering out in all directions. If he had wanted to get over the Rockies to the west, which were his orders, then he would have had to be in Hudson Bay by July when the regular brigade set off over the Rockies to Fort Vancouver, arriving in November before the worst of the winter set in. Jeffrey had an option, though. He could proceed with what was called the 'winter packet', a small company of men who went on foot, carrying their supplies on their backs, and with dogs and sledges. Effectively they walked in snowshoes over the snow, and could not take advantage of the swifter and easier method of canoeing, as the rivers were by this time frozen.

The members of the committee back in Scotland read from Jeffrey's letter:

I continued to trudge from post to post, getting a fresh man and fresh dogs at every post I came to en route. I generally remained at each station for a few days to refresh for another stage . . . arrived at Jasper House on the 21st March 1851. All this distance I walked on snowshoes, the snow being on average 2 feet deep . . . A distance of 1,200 miles. During the journey I slept with no other covering other than that found under the friendly pine for the space of 47 nights . . . several occasions the temperature standing from 30F to 40F below zero. I found no bad effects from exposure, the only thing that happened to me, was that once or twice I got slightly frost bit; that was nothing uncommon amongst us, and little cared for.

He added, for good measure, that he soon would be able to plot his own course to the Pacific. Give or take the extra miles as he zigzagged on his descent of the Rockies, he was thinking of around another 700 miles. His main concern, however, was his lack of collecting. He had been able to find some beetles and had stowed them away in his backpack, but as for other items of natural history, he discovered that he was unable to find much owing to the severity of the winter weather. Even birds were scarce, as the majority were migrants which appeared only in the summer. On the plus side, he had spent only a meagre amount of money on provisions.

Back in Edinburgh that September, almost a year to the day since Jeffrey had set foot on the Hudson Bay shoreline, the letter was duly read out, then printed and distributed. The news was posted off to inform the 281 subscribers and, just as summer came to a gentle close, the waxed seals securing the envelope must have been broken and the letter unfolded upon polished mahogany desks across the land. Many subscribers may have read it out to their nearest and dearest with a shiver, and then put it to one side with little thought except a hopeful feeling that the seeds might be a bonanza for the future. Jeffrey's letter reads as follows:

The small collections that I have made since my arrival in the country will reach Britain by the return of the Hudson's Bay

Company's ship; my journals, after this date, will likewise accompany the objects. The collections that I may make this summer will not reach England before the autumn of 1852; that is if I don't find an opportunity of sending them via Panama, at the close of the season.

My expenditure with the Hudson's Bay Company has not yet exceeded £30. I took many things from York Factory, in the way of supplies of different kinds, a whole year. When I made up my mind to start with the winter packet, I got all my surplus stock disposed of at Cumberland, at no loss. This, of course, is charged by the Hudson's Bay Company the same; but I have the money received from what I disposed of, which is available at any time.

I intend making as full a collection of 'Coniferae' of America as may be in my power. Some of the common sort I will preserve a few specimens, for such purposes adding them to the riches, which I hope will soon be, contained in the Royal Botanic Garden Museum.

I will write by the first opportunity from the west side of the mountains. I would not be surprised if this letter will reach you before this.

I have little more to add at this time I hope to have a letter from you this summer. I will reach Fort Vancouver from the north, in all likelihood, this autumn; at all events, that will be a place where my letter will be safest to be addressed to me; – there I will always find.

I am, Sir, Your humble servant,

John Jeffrey

In fact, Jeffrey had started plant collecting on the very first day after landing. Not ten miles above York Factory he had gathered seeds of *Abies nigra*. But it seemed to him but a token gesture, considering that it was now the best part of a year since he signed his contract.

On 26 April 1851, Jeffrey set off from Jasper House accompanied by chief trader Robert Clouston to cross the Rockies. If he had been apprehensive about this challenging part of the journey, he had every reason to be. Together the two men reached the limits of travelling by horse as the snow became too deep. Their local guides who were acting as carriers then declared they would not take their baggage any further. The pair shouldered their bags, slung their rifles over their shoulders and set off on snowshoes. Less than a day and a half later they reached the most northerly point of the Columbia River, called Boat Encampment, then swept down the Columbia River by canoe to Fort Colville, arriving around 12 May. Clouston proceeded on to Walla Walla and thence to Fort Vancouver, while Jeffrey was handed over to another chief trader, Alexander Caulfield Anderson.

Jeffrey used Fort Colville as a base for just a week, exploring the Kootenay River and the Pend d'Oreille region. Then he struck out on his own, after initially wandering through a considerable part of British Columbia with Anderson. It appears they parted company somewhere on the Similkameen River, and Anderson arrived at Fort Langley on the Fraser River on 15 July. If Jeffrey had been left daunted by the requirements of travel so far, there was no sign. Now alone, as far as any records show, he set off crisscrossing vast tracts of land at almost dizzying speed. Back and forth he went to Vancouver Island, no doubt seeing plants with ripening seeds, noting the exact spots, hoping against hope that he would be able to return at exactly the right moment to collect the mature seeds. Although details of Jeffrey's travels now become sketchy – he did not keep a diary like David Douglas, or indeed write copious letters as did Thomas Drummond – the only certainty is that he made confident and ambitious travels to achieve his aims, and noted down a date and location for the various plants he gathered. Far from being injured or exhausted by the thousands of miles he had already walked, it appears that the trek over the Rockies had given him confidence and supreme physical fitness. Jeffrey positively galloped from point to point.

He roamed as far north as the Thompson River collecting between the Fraser and Columbia rivers, about 100 miles from present-day Vancouver. At his most northerly point he was in the mountains east of the Fraser in latitude 50°23' where he collected *Erigeron uniflorus*. At this location he would have been around 250 miles from Vancouver. By the end of July he was on Vancouver Island. A couple of weeks later he was on the banks of the Fraser River, near Fort Langley collecting *Abies grandis*, the giant or grand fir, one of the greatest of the firs to be grown eventually in Europe. Two days later, he was at the base of Mount Baker, around 50 miles south of Vancouver and within another week he was out on Vancouver Island, at least 20 miles over the Straits of Georgia. Even at a conservative reckoning, he travelled – mainly on foot – around 500 miles in no more than a couple of months.

Back on the mainland on the banks of the Columbia River, he settled down for the winter at Fort Vancouver, sorting out his boxes of seeds and specimens and sending seeds back to Edinburgh. It was just the same as Douglas had done there twenty years before. One of the friends with whom he passed many wandering winter hours was young James R. Anderson, the son of chief trader Anderson with whom he had walked earlier. Although James, who later also became a botanist, was but ten years old their meeting was of huge significance to Jeffrey's story. But half a century was to pass before that relevance was revealed.

Later in the winter of 1852, Jeffrey went island-hopping both to the north of Fort Victoria on the Hudson's Bay Company's ss *Beaver*, and to the San Juan Islands, previously explored by the crew of the *Discovery*, led by Captain Vancouver in 1794. In the spring he returned again and gathered a new form of flowering currant, the species that had been such a success for the Horticultural Society of London when brought back by David Douglas. Jeffrey's variety was eventually named *Ribes lobbii*. He also gathered the juniper and an anemone but his most important find was back on Vancouver Island on 24 April. It was found to be a new introduction into Britain and was eventually named *Tsuga heterophylla Sarg* (western hemlock), although Jeffrey had named it *Abies taxifolia*.

The entire summer of that year he followed exactly the plans which he had previously expressed to chief factor James Douglas.

Douglas had written to chief factor John Bellenden at Fort Vancouver telling him that Jeffrey was 'on the eve of proceeding on his professional pursuits by way of Nisqually and Cowlitz' on his way to Fort Vancouver down on the banks of the Columbia River, and that he would be required to request a cash advance from the money lodged by the Oregon Association back in Edinburgh with the Hudson's Bay Company for the means to travel on to California.

Jeffrey did exactly this. He went south by the Williamette Valley, in the neighbourhood of Puget Sound (named after one of the officers on Vancouver's *Discovery*), climbing high up the Cascade Mountains, on the border between present-day Washington State and British Columbia's western edge, stamping through the area that David Douglas had explored, and on through the Umpqua country. On his way, he discovered various delights including a lily which he named after the Umpqua country in which it was collected. He never found out that another explorer named Kellogg was to collect it just two years later and within a decade the lily had been renamed *Lilium washingtonianum*.

He ranged over Mount Shasta, discovering a new pine which he named after Professor Balfour, before moving on to the Salmon Mountains and Trinity Mountains. Here, he also collected seeds of various trees: *Pinus monticola Doug.*, *Tsuga mertensiana*, *Abies magnifica Murr.* (no doubt named after Murray back in Edinburgh), *Pinus contorta Doug. v. latifolia* (the beach or shore pine), *Pinus jeffreyi Balf.* (the black or Jeffrey pine) and *Libocedrus decurrens Torr.*, the highly aromatic incense cedar tree. For good measure he also climbed Mount Jefferson before retreating to Fort Vancouver to hole up again during the winter, and sort out his collections. Conscientiously, he sorted all his finds, and in the time-honoured, traditional way of all plant collectors, numbered each plant, seed bag and box before dispatching them. Box five was dispatched on 22 January 1853, box six on 15 February and another box of duplicates, some of which were sent by post, some overland, back the route that he had trudged, to York Factory, and yet others by ship around Cape Horn, totalling in all nine packages. Three of the boxes never reached Edinburgh. By coincidence, in Edinburgh that January, the chief factor of Fort Vancouver, John Bellenden, was visiting Scotland, so he was able to report at first hand to the committee meeting of the Oregon Association that he considered Jeffrey to be a very hard-working, 'energetic and industrious person, and that he was much thought on by all who had seen him'.

On 6 April 1853, the same day that he received $500 from the Hudson's Bay Company, Jeffrey set off southwards as arranged. He veered west to the Sierra Nevada Range in August and September. By the first of October he discovered a *Cupressus*, which he named *macnabiana* after the gardener at the Edinburgh Botanic Garden who had recommended him so highly. One week later he drew a bill to the value of £200. On 20 November, he was in San Francisco, and was apparently ill there for a few weeks, although he did manage to dispatch his tenth box around the beginning of January 1854. Then John Jeffrey, aged twenty-eight, simply vanished. He who had been so fit, energetic, scrupulous in observing protocol, and avoiding all conflict with Hudson's Bay employees, simply disappeared from the face of the earth.

Soon after he left San Francisco a letter arrived for him from the Oregon Association in Edinburgh. Living in San Francisco at that time was the brother of Andrew Murray who was the secretary of the Oregon Association, and Jeffrey's letter arrived at his office. In his response to the committee in Edinburgh, Murray replied:

> I yesterday received your letter enclosing one for Jeffrey and I went again to McKinley, Garrioch & Co., as they have deciphered his address to be Fort Yuma, on the Gila River (close to where it joins the Colorado) of where he says he will probably be until the first of August, and directed letters to be forwarded by Adams and Co.'s Express to the care of their agents at San Diego, Mr F. Ames.
>
> They, McKinley, Garrioch & Co., say he is a hard-working, enthusiastic, very steady and temperate man [teetotal], and that just before starting for San Diego, he was some three weeks arranging the proceeds of his excursions, and they doubt not that

he dispatched them. He had been for some weeks sick before that, which accounts for passing his long stay in San Francisco.

Various theories and rumours filtered out over the years. One theory was that he joined an American expedition which left San Francisco that spring for Fort Yuma to explore the Gila and Colorado Rivers, which seems to tie up with Murray's letter. One of the more colourful stories was that he had somehow been swept up in the gold rush. Certainly, on 24 August 1852, it had been noted in Edinburgh that the third box of the plants and seeds he had sent came from San Francisco (the postage charge being an astonishing £133 which luckily the committee managed to have waived by the post office in Edinburgh) and also contained small specimens of gold from Queen Charlotte Islands, close to Vancouver. However, there is no other evidence.

The Association were none the wiser about Jeffrey's disappearance. As ever, they remained divided between being entirely satisfied and extremely dissatisfied with his work. Many of his boxes were lost en route home, so judging him by the quantity of seeds and finds was impossible. The fact is that by September 1853, in the last autumn of Jeffrey's life and with his contract drawing to a close within two months, there had also been several critical letters from subscribers. The following spring, in March 1854, Andrew Murray, secretary of the Oregon Association, noted that although another box of seeds had arrived, they were disappointing in both quality and quantity, and Jeffrey was not to be offered another contract.

That the Association were going to be disappointed was almost a foregone conclusion. When David Douglas first visited the area in 1825, fewer than 400 Europeans inhabited the vast tract of country through which he had travelled. Twenty-five years later, the area was well documented, well travelled and picked over by various seed collectors. It was by no means virgin territory, and no longer so inaccessible. Twenty years before, Douglas had made it plain that he could not collect all the seeds and plants from the countryside sometimes owing to being there at the wrong time of the year.

Douglas also knew only too well that many seeds had been left to rot in London, and of course that others had not survived the journey. So Jeffrey had followed Douglas' footsteps as he would have been fairly certain that he would find plants which Douglas had missed, or which had been lost in transit, had died, or been allowed to rot. Douglas' reports on these areas pointed to further treasures there for the taking and Jeffrey's instructions were to find them, rather than to set off on unknown trails.

Douglas had died in 1834, and that same year Thomas Nuttall from Harvard University abandoned his academic career to accompany Nathaniel Wyeth, picking up specimens along the Columbia River. This was followed by an official United States Exploration Expedition under Charles Wilkes, which visited the Pacific North-west, scouring the land from Fort Nisqually east to Fort Colville and as far south as Fort Vancouver, then finally through the Williamette Valley to their ship at San Francisco. Around 1,218 specimens found their way to the US National Museum in Washington, DC. Compared with this record, Jeffrey produced a creditable 400 species which not only were given mention in the history books, but also were lodged eventually in the herbarium of the Royal Botanic Garden in Edinburgh. He certainly found significant quantities of conifers, which were after all the main point of his expedition.

The news of his disappearance finally filtered back to the good burghers in Edinburgh, after it became apparent that no letters had ever been collected at San Diego, and when no packages arrived back in Edinburgh, and no one met anyone who reported seeing him. It was left to Andrew Murray, the secretary of the Association, to issue a mild rebuke to the critics, and no doubt to guard both his own reputation, and that of his brother in San Francisco:

> Some subscribers to the association, remembering totally the third and last year of Jeffrey's engagement terminated unsuccessfully, and that they had just reason to be seen dissatisfied with his conduct during that year sometimes speak of his expedition as the failure. But it is unjust so to term it; and if they would only

remember the quantities of novelties which were discovered and introduced through his means, they would rather treat it as a great success, which only assumes the aspect of a partial failure from the knowledge that, great as it was, it ought to have been, and might have been, greater still. No one could have worked more conscientiously and more perseveringly than Jeffrey did during the first two years of his employment, and bearing in mind that that Menzies and Douglas went to virgin country, his collections do him credit, even as compared with theirs.

One man who recalled Jeffrey was James Anderson, son of the chief trader Alexander Caulfield Anderson, who had guided Jeffrey from Fort Colville to Fort Victoria on his overland journey during the winter of 1851. Although he was only ten years old at the time, he remembered Jeffrey well and one incident in particular.

Few living people possibly remember to have met Jeffrey, the naturalist, after whom many native plants are named. Mr Jeffrey reached Fort Victoria in 1851 having come through Fort Colville where my father was stationed, and I was therefore the chosen companion of Mr Jeffrey in his nearby excursions. A woodpecker slain by Mr Jeffrey in the edge of the woods where the city nursery now stands remains impressed on my memory. I never heard anything further of Mr Jeffrey after his departure until 1911 when visiting the Royal Botanic Gardens in Edinburgh and conversing with the director Professor Balfour the name of Mr Jeffrey came up, and I was then informed by Mr Balfour that after leaving Fort Victoria he found his way to San Francisco, then in the throes of the gold excitements, and was never heard of afterwards, probably murdered by the lawless ruffians who congregate at all mining centres. Naturally Professor Balfour was greatly interested in meeting someone who actually had seen Mr Jeffrey.

After I had penned the foregoing, I came across the following, written by my late father in his essay in 1872, regarding Mr Jeffrey as follows: 'The late Mr Jeffrey, a botanist, who visited this country under the auspices of the Hudson's Bay Company, employed by the Duke of Buccleuch and other gentlemen to make collections.

Poor Jeffrey it may be added, after wandering sometime in company with the writer through a considerable portion of British Columbia and braving all its fabulous dangers, met his fate in New Mexico in 1852. He was murdered by a Spanish outcast for his mules and his scanty travelling appointments.'

So Anderson senior must have heard some rumour near the time of Jeffrey's disappearance as to his fate, and with this tantalising tale, the closest contemporary account, Jeffrey's death remains an unsolved riddle. So another explorer of the Pacific West vanished in his prime, but one who did survive, an amiable naval doctor from Auchinblae in Kincardineshire, not only trod in their footsteps, but also helped trace out the boundary between Canada and North America, along the 49th Parallel.

CHAPTER 12

David Lyall (1817–1895)

THE ARCTIC, THE ANTARCTIC AND THE 49TH PARALLEL

James Clark Ross, commander in 1839 OF HMS *Terror* and HMS *Erebus*, was one of the most renowned polar experts in the world, who had already located the magnetic North Pole in 1831. He had spent seventeen of the last twenty years within the Arctic, surviving eight winters, and now he was destined to locate the magnetic South Pole. As he surveyed the two dumpy, middle-aged ships now being refitted to carry polar explorers, he might well not have regarded them as enviable beauties. Ross himself was a hero in his own lifetime, and if that wasn't difficult enough to deal with, he was also rumoured to be the most handsome man in the British Navy: quite a reputation to live up to. The voyage to the Antarctic was planned to last four years but both ships had already experienced life to the full.

Terror had seen action in 1812 in the Crimea; suffering damage near Lisbon, she was repaired and then left inactive for a few years. When she emerged again, it was to be sent off to the Hudson Bay, northern Canada, with the intention of pushing through to Repulse Bay. The mission was a forlorn hope which not only comprehensively failed, but the ship also suffered the indignity of being shoved forty feet up a cliff by ice. All she could do was wait either to be crushed or to survive. The fates smiled, and when the ice melted, she was able to crawl back to Ireland to be once more given a major rebuilding. Having been designed as 'bomb ships', for their powerful 3-ton guns to bombard onshore enemy targets, both *Terror* and *Erebus* were built to have Herculean strength and were far from ready for the scrapheap.

At the time, ships built for the icy conditions of whaling or sealing were few in number, and these stocky bomb ships were the best available. Extra oak beams were inserted fore and aft to absorb the strains of crushing ice, the hulls double-planked and the keels sheathed with substantial copper plates. Everything which could be strengthened was strengthened, even to the extent of triple-thickness sails. Young David Lyall, approaching her fresh from medical studies in Aberdeen, would have been only too familiar with sails.

Auchinblae, Lyall's home town, and the east coast of Kincardineshire and Aberdeenshire produced copious quantities of rope and sailcloth in the early to mid eighteenth century. The women of the coastal village of St Cyrus had a strange and curious way of weaving flax for sailcloth, coiling the flax around their waists and twisting it a couple of threads at a time, one in each hand, as they walked slowly away from a helper who would be turning it at the end of the room.

Hailing from the sturdy farming parish of Fordoun, Kincardineshire, David Lyall, born in 1817, was more than likely the first member of his family to attend university, or at any rate the very first to obtain a medical degree. As was the case in medical study, he would be fully qualified by the time he was twenty-one or -two, and indeed by that time he was also licensed to practise surgery. Lyall was born into a family of cautious entrepreneurial strain, in a largely fertile area with strong arable production. A century later the rolling land and villages such as Auchinblae were made famous by Lewis Grassic Gibbon's *A Scots Quair*, which depicted small-town and farming life in the area. The rich, red soil of the land, or 'parks' as they were locally known, was the soil on which modest fortunes were made. The climate was not quite as harsh and exposed as other areas of Scotland, and over hundreds of years the soil had been worked to a fine tilth, and was therefore able to support high-yielding crops and a multitude of wild flora.

Lyall's grandfather, William Lyall of Wattieson, a tenant farmer at the east end of the parish, was noted for his pioneering experiments with turnips as a crop. The year was 1756, and the introduction of this humble turnip was to revolutionise farming. Until then there was no crop that could be reliably produced on Scottish land, successfully lifted out of the ground, and stored through the winter in such quantities as to be able to keep the cattle well fed and alive. Most farm animals, except for a few essential breeding animals, had to be slaughtered each autumn. The remaining cattle, carefully chosen for the future of their bloodlines, were so weakened by springtime that they usually had to be carried out into the new fields. This process was evocatively known as 'the lifting'. Within twenty years turnips were under general cultivation in Scotland, but pioneering William Lyall, having begun on a very modest scale, was able to sell his crop at the rate of one penny per stone, and both his pocketbook and his reputation were enhanced. Although his success and far-sighted agricultural gamble had paid off, he remained only a tenant farmer, and his son Charles left working the land, although not the immediate area.

He moved a few miles down the road to Auchinblae, and was eventually the proud co-owner of a mill in partnership with a Mr Kinnear, and lived in the town until his death. Charles' wife Elizabeth was still alive at the time of the 1851 census. She must have been around thirty-eight when their son David was born. With an agricultural improver as a grandfather, an astute businessman as a father and academic hopes probably firmly pinned on him, young David did not let them down. He proved to be a first-class pupil in botany, and studied medicine at the University of Aberdeen where he gained his qualification to practise surgery. Life was set fair.

Living not far from the port of Montrose, where seafaring had provided vital and lucrative trade with the Low Countries for a couple of centuries, and indeed seen the creation of town houses of great sophistication, and having spent his formative years in Aberdeen, where the harbour throve, he would have been familiar with the sea. Like so many before him, his answer to finding a career that would show him more of the world was to apply to be a naval surgeon. Either Lyall had a charismatic personality, or he was more than usually determined, for this dream job aboard HMS *Terror* as assistant surgeon was in fact his very first naval appointment. He was not, however, a complete novice, having already spent some time on board a whaling ship in the Greenland area. No doubt his collection of botanical species from Greenland, where he would already have gained experience in extreme and cold conditions, combined with his medical degree, gave him an edge over other candidates.

There is no record of how many applied for the post, but it would have been very unusual if a queue of excited young men had not formed. The aim of the expedition was to find the magnetic South Pole, and this was to be part of Britain's contribution to an international year of co-operation whereby European nations would set up magnetic observatories around the world, co-ordinating readings on fixed dates and comparing readings. Expeditions from France and the United States were also hell-bent on exploring the area. Behind a façade of well-meaning international co-operation these expeditions represented, if not an outright race, at least an ambitious jostling to establish who could gather as much scientific information as possible, establishing supremacy of intellect if not ownership of land. Captain Ross had a second ambition: he wanted to plant on the South Pole the flag he had already taken to the North Pole. For Lyall, to play a part in such an adventure was a notable achievement, especially as his equivalent appointment on board *Erebus* was Joseph Hooker.

Joseph Hooker was the brilliant, cherished, determined and hothoused son of William Hooker, Professor of Botany at the University of Glasgow and a networker and facilitator of note, who had already mentored many young explorers. As his son Joseph

prepared to sail on HMS *Erebus* in 1839, he was still in his post at
Glasgow University, and supervisor of the Glasgow Botanic Garden.
Joseph, the younger son of William and his wife Maria Dawson,
was born on 20 June 1817, just nineteen days after David Lyall. Both
qualified in medicine at Scottish universities in the same year, 1839.
That, however, is the full extent of their similarities.

Educated at Glasgow High School and later at Glasgow
University, from the tender age of seven Joseph Hooker had sat
in on his father's botany lectures. When his father received letters
from David Douglas in far-flung areas of north-west America and
Canada, he listened intently to special messages just for himself.
When Douglas returned home in 1824 he visited the Hooker family
in Glasgow. So, from a very early age, Hooker was surrounded by
and intrigued by travellers' tales. In later life, he recalled sitting on
his grandfather's knee, looking at the pictures in Captain Cook's
Voyages. He was particularly struck by one of Cook's sailors killing
penguins on Kerguelen's Land, on Antarctica, and he remembered
thinking, 'I should be the happiest boy alive if ever I would see that
wonderful arched rock, and knock penguins on the head.'

However, pleased as Hooker was to be appointed as assistant
surgeon on HMS *Erebus*, he was not entirely satisfied. Ross was
a friend of his father's, and had been instrumental in assisting
Joseph's appointment. Before they set sail, an eminent geologist
Charles Lyell, from Kinnordy – with a similar name but unrelated
to David Lyall – had sent William Hooker the proofs of Charles
Darwin's *Voyage of the Beagle* and he fretted about his ability to
match Darwin's success as a naturalist. Hooker was not alone in
his reservations about his own relative merit. Ross himself would
have preferred to have had a naturalist of the calibre of Darwin on
board, and certainly felt that the youthful Hooker was unproven.
In appointing him to the fairly lowly position of assistant surgeon,
Ross was hedging his bets. While he encouraged Hooker in his
botanical quests on the voyage, he could justify his presence on
board by having him fill the necessary medical post. Darwin's
presence on the *Beagle* had been under very different circumstances.

He had been financed by his Wedgwood uncle, so his role was closer to that of a paying passenger than as a paid member of the crew. Joseph observed testily to his father, who was unable to finance him on the Ross expedition, that Darwin had also been inexperienced before his trip: 'What was Mr Darwin before he set out? He, I daresay, knew his subject better than I now do, but did the world know him? The voyage with Fitzroy [on the *Beagle*] was the making of him.' Joseph certainly hoped that this expedition would set him on a similar illustrious path. Lyall, on the other hand, never recounted how he felt.

The expedition set off in the autumn of 1839. Both ships had a crew of sixty-four, and both were packed with supplies to last for three years. On board were copious quantities of food to counteract scurvy: carrots, pickled cabbage and vegetable soup. For good measure there was a small flock of sheep. Alongside this were ice saws, portable forges, and generous quantities of clothing suitable for extremely cold conditions. As the well-equipped ships cruised down the Channel, Ross was able to exclaim that 'It is not easy to describe the joy and light-heartedness we all felt, as we passed the entrance to the English Channel, bounding before a favourable breeze over the blue waves of the ocean, fairly embarked on the enterprise we had all so long desired to commence.'

His crew, crowded into fully packed vessels, each of which had a deliberately shallow draught (11 feet) so that it could slide close to shore, may not have felt quite so euphoric. The dumpy ships wallowed in the ocean rather than skimming the waves, and had an extremely uncomfortable movement when sailing.

The two ships took all of twelve months to go from England via Hobartstown (now Hobart) in Tasmania, to commence the first section of their Antarctic expedition. Ross then acted on a tip-off from a sealing ship that there was a stretch of open water beyond the pack ice. It being late autumn of 1840, the Antarctic was now on the threshold between spring and summer and the pack ice was starting to open. Ross aimed for the 180th line of longitude and then turned sharply southwards, which conveniently enabled the ships to remain undetected from the rival American and French expeditions of d'Urville and Wilkes.

On Christmas Day 1840, the first iceberg was sighted and on New Year's Day 1841, the ships crossed into the Antarctic Circle. Ross deemed it to be officially cold, so the crew were assigned their thick winter clothing of woollen felted garments. Three days later the pack ice loomed and in what would seem to modern sailors an amazing fit of seamanship, Ross ordered the crew of the *Erebus* to ram repeatedly the exact same point in the ice, which he hoped was the weakest section. For ships powered only by sail, this by any standards was astonishing. Although the rumoured 'lagoon' of open water was not immediately apparent, the two ships could now edge their way to what they hoped was the entrance to the southernmost point, thereby reaching the South Pole. At last the sheer strength of the reinforced ships paid dividends. Within one week they had broken through the pack ice into the lagoon, and knew that they were in an area never before visited by man. As they watched their compass they could see that they were drawing ever closer to the magnetic South Pole. By 11 January, they had sighted land. As there was no evidence of human habitation, they concluded that ownership was not an issue, and set about naming various areas of water and the landmass which they could now see in the distance.

Now known as the Ross Sea, this great inland lagoon, the rough 'bite' in the circular shape of the Antarctic landmass, enabled them to sail much closer to the magnetic South Pole. They named the range of mountains Admiralty Range, and Victoria Land after the newly crowned queen, and what seemed to them most incredibly an active volcano, sighted at an estimated distance of 100 miles, they named Mount Erebus, after one of their ships. For good measure another volcano was named after the ship upon which Lyall was serving, Mount Terror.

Within a month, Ross ordered the ships to leave the Antarctic following the route by which they had entered the area. Having spent four years in the Arctic, Ross had acquired a wealth of experience, and knew only too well that the pack ice could easily close in again and entrap them without warning. It was by now 9 February and the short Antarctic summer was coming to an end. They returned to Hobart in April 1841, for recuperation and a major refit which took several months. For Lyall and Hooker, this meant they could take their time to search for botanical treasures. In the autumn of 1841 and again in late 1842, the two ships repeated the inaugural voyage into the Antarctic Ocean, and although productive from the point of view of collecting more scientific data, they never again reached quite so far south. It would be fifty years before anyone would venture to the point they reached on the first year's expedition. Nobody ever did it again by sail.

On 4 September 1843, they returned to England, laden with copious data on oceanography, as well as botanical and ornithological collections for which Lyall and Hooker were largely responsible. Although most of their specimens from the Antarctic areas were algae, it was the plants and seeds they had collected mainly from New Zealand which intrigued Sir William Hooker and the skilled nurserymen in Britain. Ross received a knighthood, and Lyall now had an established reputation as a botanist. He could look with some pride upon the enormous botanical collections to which he would have contributed substantially. He could also bask in his recommendation from the Admiralty as 'meriting the highest commendations'. Almost as important, he had made, in Joseph Hooker, a friend for life.

During the winter months, Lyall and Hooker had collected interesting algae from such places as the little-known island of Kerguelen, which lies in the Indian Ocean just to the north of the Antarctic landmass. Here Lyall found a plant eventually named in his honour, *Lyallia kerguelensis*. Obscure as this might seem today, Victorians delighted in discoveries of mosses, algae, grasses, beetles, fungi, stuffed birds, relics of 'native' peoples, birds' eggs by the dozen, butterflies and anything of geological interest. Lyall had amassed a substantial collection on his very first expedition. After marking time, serving as an assistant surgeon for five years with another naval ship mainly employed in the Mediterranean, a chance came his way. This chance came not only through his friendship with Joseph Hooker, which kept him directly in touch with his father William Hooker who was by now Director at Kew, but also Lyall's growing reputation must have contributed to his next appointment as surgeon and naturalist on board HMS *Acheron*. Under Captain Stokes, the *Acheron* was to survey the coastline of New Zealand. Here Lyall found one of the most impressive of all buttercups, with outsized white flowers. New Zealand shepherds called it a water lily, but its official name now is *Ranunculus lyallii*. Lyall had just time to pack his warm-weather clothing away when another highly desirable appointment came his way in 1852.

On board his old ships *Terror* and *Erebus*, Sir John Franklin had disappeared, along with his entire crew, in the Arctic while searching for the North Pole and attempting to find the elusive North-west Passage connecting the tip of eastern Canada with the tip of Alaska, thus effecting a short cut from one ocean to another without the tedious sail round the treacherous tip of South America. Although Franklin's expedition had been one of the best equipped ever to set out, both ships and their crews seemed simply to have vanished. As this disappearance had been so publicly reported, and had caught the imagination of the British people, multitudes of search parties were sent out on various ships. Attached to one of these expeditions, Lyall found himself appointed as surgeon and naturalist on board the *Assistance* under Captain Belcher. Lyall was also elevated to the role of lieutenant, and given charge of one of the sled teams. The search parties from his expedition failed to find any trace of the Franklin party, but they did find algae and mosses. Although of huge and popular scientific interest in Victorian Britain, they were of scant use for the suburban gardens back home.

If he thought that a more comfortable naval appointment might be forthcoming, Lyall was in for a rude surprise. His career so far might well have taken him to far-flung places and given him opportunities to assemble plant collections. There was one aspect of the navy which he had so far avoided. For sixteen years he had not been on board a ship that had fired a gun in anger. This peaceful life came to an abrupt halt when, in 1855, he was catapulted from the relatively sedate life of a peacetime naval surgeon and botanist into the full fury of naval warfare in the Crimean War. Surviving its horrors, Lyall was then sent off to a very peaceful and fascinating task. It was 1858, and the border between Canada (owned by Great Britain) and the USA, along the 49th parallel, had to be delineated all the way from the Pacific coast to the summit of the Rocky Mountains. Although the 49th parallel runs in an 'organised' straight line right across the continent, at the western edge where it hits the Pacific Ocean there is a complex area which dips down below the Strait of Juan de Fuca, and including part of Vancouver Island in Canada. Lyall, with Colonel Sir John Hawkins of the Royal Engineers, was detailed to survey the 49th parallel.

Stationed and based in Victoria, on Vancouver Island, Lyall would have found it a far cry from the early days of Captain Vancouver and Archibald Menzies in the 1790s, and huge changes had taken place even from the time of John Jeffrey six years before. Lyall spent three years in the area, not only charting the boundary, but also serving as 'surgeon and naturalist' to the Boundary Commission. For assistance, he had Sapper John Buttle of the Royal Engineers, who, in order to be prepared for this task, had received training in botany and horticulture at Kew Gardens. Vancouver Island may have seemed civilised to those stationed there, but less so was the scene on the mainland. Until that summer of 1858, no more than 192 European women, mainly the wives of settlers, occupied the western area of Canada, an area the size of Western Europe. Now, however, there was an influx of tens of thousands of gold-seekers, rushing up the Fraser River to pan for gold. Although this was happening just over the Straits from Vancouver Island, little appears to have ruffled the calm island life.

Lieutenant Samuel Anderson of the Royal Engineers reported on 28 March 1860:

In Victoria, I used to get up about nine, read the newspapers, take a few Solar observations with a sextant till 12, have luncheon, and ride up to town about 2 p.m., lounge about the town paying visits and shopping till three, then go for a ride, get home [to the officers' mess] about 5.30 p.m., have dinner at six o'clock, a cup of tea at 7.30 p.m., rubber of whist (for love) [as opposed to money] till eleven, and then turn in and that was our ordinary employment. We used to be overrun at various portions of the day by naval officers coming onshore for fun, and in the evening we used sometimes to have as many as a dozen at a time in our messroom, and we were all great friends with them.

Into this humdrum existence came 'our surgeon, a Dr Lyall, Royal Navy of Aberdeen,' continued Lt Anderson, 'who is a most experienced man. In addition to having been in every ordinary portion of the world, he has been on an Arctic expedition and on an Antarctic expedition and though not a very talkative man, we get curious yarns from him at times.'

By now Lyall was in his forties and used to life on a deck with the occasional sortie ashore, but he was about to exchange Victoria's gentle way of life for that of a mountaineer. He had been asked to scale the Rockies, a challenging climb from sea level to the 8,000-foot summits, in order to define the cut-off points between the ever enlarging United States of America, and Canada. As the party traversed the 49th parallel, Lyall reported meticulously not only on the plant life, but also the food requirements for the men of the expedition, as well as the horses and mules.

As soon as the eastern slope of the Cascade is contained, all difficulty about fodder for animals ceases, and parties may travel from thence to the Rocky Mountains without grain for beasts of burden.

At the same time, with a large organized party like ours, where the mules were kept constantly on the move as long as the ground could be travelled over on snow . . . it was considered advisable to pack a certain quantity of grain for the animals, in order to preserve them in full strength and vigour.

A sailor and physician he may have been all his adult life, but he still possessed the sharp, observant eye for more down-to-earth agricultural plants and habitats, learned from a rural childhood.

At a height of 6,480 feet, at the camp of the astronomical station on the Rocky Mountains close to the 49th parallel, they climbed up another 2,000 feet where Lyall harvested the rich vegetation growing on what must have felt like the top of the world. By collecting plants on the summits as well as at sea level, around the immediate vicinity of Victoria on Vancouver Island, Lyall managed to accrue a vast and rounded collection. He also had a keen knowledge of their uses and properties. Cedar trees, he noted, were an exceedingly useful source for the local Native Americans, just as David Douglas had observed thirty years before in his diaries. 'The trunk is used to form their canoes, and then split into slabs [or planks] which is done very easily, to build their permanent huts or lodges. As for the stringy bark and the integuments of the root, these are plaited into useful and ornamental articles of clothing and household utensils.' He noted that these cedars and the Douglas firs were the most useful trees from this part of the coast, and were now part of a large and increasing export business, both for the spars on ships and as ordinary lumber.

Within fifty years, Menzies, Vancouver and Douglas had effectively stamped their names on the area. The town of Vancouver was growing in size, and the *Pseudotsuga menziesii*, named in Latin after Menzies, was by now also commonly known as the Douglas fir. The boundary was now officially mapped, and in a grand finale to what appeared to have been a successful and golden career, Lyall returned with a massive herbarium amounting to 1,375 species. This so pleased Sir William Hooker that pressure was brought to bear for Lyall to be appointed to an official naval job as staff surgeon at Woolwich. This was a post which would leave plenty of time for naming, cataloguing and distributing his collections, which now amounted to a colossal 6,700 specimens.

Lyall covered more extremes of botany than almost any other explorer, from the Arctic to the Antarctic, and across oceans from

New Zealand to the west coast of North America. He brought back thousands of plants, many of which were of great garden popularity. To Lyall we are indebted for the plants of New Zealand: curious little daisy-flowered *Celmisias*, several varieties of *Hebe*, that spiky plant beloved of park garderers, and *Phormium*, which is now so popular it is to be seen growing in gardens across Europe. His *Ranunculus lyallii*, the huge giant buttercup with its spectacular flowers, is one of the very few named in his honour. From his treks along the 49th parallel in north-west America came two of the prettiest of dogwoods, *Cornus canadense*, a creeping version, and *Cornus nuttallii*, the white-flowered mountain dogwood. From that area too came the *Lilium canadense*, a wild yellow lily, *Tellima grandiflora*, with its rich bronze leaves, and *Trillium grandiflorum*, one of the largest of the appealing three-leaved *Trillium* which is commonly known as 'wake robin' and carries extravagant white flowers.

Back home again, and awash with plants, Lyall enjoyed a few happy years completing his report at Kew, and was eventually elected as a member and fellow of the prestigious Linnean Society in 1862. There were to be no more sorties abroad. Within a few years he had completed his final naval appointment as surgeon for the Pembroke dockyard, and it must have been from there that he met the daughter of the local doctor, Dr Rowe of Haverfordwest, whom he married in 1866, having three children and retiring to Cheltenham. The citizens of that town never appeared to have an inkling of the 'curious yarns' he could have told. Lyall appears to have merged into the middle classes of a middle England town and effectively retired from life. When he died in 1895, his old friend Joseph Hooker wrote his obituary, covering a remarkable life which had swung from Auchinblae and Aberdeen to both the Arctic and Antarctic, New Zealand and the west of America and Canada.

CHAPTER 13

Thomas Thomson (1817–1878)

CAPTURED IN AFGHANISTAN

Thomas Thomson's obituary ran to eight pages in the *Proceedings of the Royal Geographical Society 1878*; with typical British understatement, less than half a page is devoted to an ordeal about which dozens of books have been written, and films made. It reads as follows:

> At Ghuznee [Ghazni], Afghanistan [about 100 miles from present day Kabul] he was attached to the 27th Regiment of Native Infantry, and had his first attack of fever, soon to be followed by the horrors of Afghan campaign, of which he was one of the few survivors. Very soon after he had quitted Cabool [*sic*], the detachment left there was destroyed, excepting a few of its officers and men who fled to Ghuznee, where, along with Thomson's detachment, they were beleaguered during the winter of 1841/42, and where, after daily losses of their comrades by the rules of our sickness, starvation, and the enemy fire, they capitulated, to be subsequently imprisoned by their treacherous and savage captors. From Ghuznee, Dr Thompson and his fellow prisoners were afterwards sent to Cabool, and from thence were being transported to Bokhara to be sold into slavery; but on their arrival at Bameean they bribed their captors – for a

ransom of £2,000 and the promise of a pension for life from the British government – to conduct them back to the advancing army of relief.

Back in Glasgow, the Thomson family must have had few illusions about the outcome of the war. Horrified readers were clustering round reports of wholesale massacre, of an entire section of the British Army in India being wiped out in the mountains of Afghanistan by mere tribesmen. How could this have happened? First there was shock, then outrage as more and more details began to emerge, although the reports only ever revealed part of the story. What hit the pages, written carefully for an anxious British readership, was that the heroic British Army, assisted by loyal(ish) 'native' recruits, had been set upon in a cowardly fashion – in other words, ambushed – and virtually annihilated. Probably, the Thomson family dug out their black mourning clothes and made themselves ready for the worst. What a blow it would be for the family's hopes, and years of glowing pride, as their eldest son climbed almost effortlessly up the academic ladder.

Thomas' father had risen, through sheer brain power, from the minor parish school of Crieff, Perthshire, to the post of Regis Professor of the Glasgow Philosophical Society. In 1816, he married Miss Agnes Colquhoun, the daughter of a Stirling distiller, and the following year their son Thomas was born. Young Thomas demonstrated a precocious intelligence and talent for sciences. His academic ability mirrored that of his father. But where his father had had to struggle up the academic ladder, Thomas, if not quite born with a silver spoon in his mouth, had at least arrived into a highly nurturing home, where academic opportunities surrounded him. His schooldays were spent at the acclaimed High School of Glasgow, where he showed a natural aptitude for classics and mathematics, but a much greater interest in physical and biological sciences.

There was just a whiff of the child prodigy about Thomson junior. Luckily, however, he had one special classmate of equal intellect,

Joseph Hooker, who was to become a lifelong friend. Joseph's father William was also a friend of Thomas' father, and was an equal in academic ability. In fact, the families' paths crossed constantly both in Glasgow and on the other side of the world.

Young Thomson was therefore in the constant company of acute and stimulating scientific minds and was credited with discovering, at the tender age of seventeen, that the fossil molluscs at Dalmuir on the Firth of Clyde, were not deposited, as previously thought, during a glacial age, but during the late Tertiary period when the banks of the River Clyde, at least as far inland as Glasgow, were submerged under the sea. His father, however, pressed him to concentrate on chemistry, even sending him to study for a winter at Giessen where he learned under one of the most advanced thinkers of his day, Professor Liebig. The professor apparently regarded Thomas as one of his most promising students and the pair of them went on to isolate pectin acid in carrots.

But there was more than dazzling intelligence to Thomas Thomson. Far from being a pasty-faced, academic recluse, he was an enthusiastic and athletic mountaineer and, unusually for his era, a daring rock climber. When he decided to study medicine, he came under the spell of his best friend's father, William Hooker, who was a professor of botany and a family friend. Thomson swivelled his considerable intellect towards botany, and was determined to make this his life's interest. In this he was encouraged greatly by Joseph Hooker's father, William. This friendship was further cemented in the summer of 1838 when Joseph and Thomas set out for a summer of Scottish geologising.

Thomas must have impressed Hooker senior, as it was with his blessing that he set off as an assistant surgeon for the East India Company. Curiously, on his arrival at Calcutta he was appointed, not as a surgeon to the company, but redirected to be curator of the Museum of the Asiatic Society. Although there is no concrete evidence, we can only assume that the hand of Hooker influenced his recommendation for the appointment. Life appeared set fair. However, despite flinging himself into the task of arranging the fine collection of mineral specimens, he had scarcely time to catalogue more than a handful before he was wrenched out of his quiet life by a command from the army, to be sent to Afghanistan with a party of recruits who, like himself, had just arrived from England. More than likely these raw new recruits were destined to fight in the British Indian Army. From Calcutta the party marched across the continent.

This long march to Afghanistan took from August 1840 to June 1841. In the far distance at Kabul, it was the calm before the darkening clouds of winter and the invisible but menacing tribesmen closing in. Thomson roamed around botanising among flora that had been previously unknown and unrecorded in Europe. Added to this were forays to study geology. Little did he suspect that while carefully drawing, recording and pressing flowers, imagining himself securely protected by the 27th Regiment of Infantry, he would be one of the very, very few to survive the following few months. He was about to be caught up in one of the most vicious battles of the interminable Afghan wars.

This erupted as a result of the British installation of a puppet ruler, Shah Shuja, in the capital of Afghanistan, replacing Dost Mohammed. To establish the British choice of ruler, a massive military contingent which later became known as the Army of the Indus was assembled. It was a not insignificant undertaking, with 18,000 soldiers, plus some 38,000 camp followers and 30,000 camels. It appeared to occur to nobody that the camels were the wrong breed of animal transport for such an adventure. This vast caravan set off in considerable style: one section coming from the south at the Sind, another from Bombay. Army officers were accustomed to travelling with all the comforts of home plus plenty of miscellaneous items to help pass away the time after delivering what they fully expected to be a short sharp shock to the local peasants. The officers' cigars alone required two camels to transport them.

Little wonder that the entire exercise was christened the 'Great Game' by future historians. One officer of the 16th Lancers had

forty personal servants with him, and a brigadier felt it necessary
to equip himself with preserved food, eau de cologne, and all
the accoutrements for leisure pursuits such as polo. For good
measure, a pack of foxhounds came along too. Unfortunately, some
preparations were soon found sorely lacking. The oppressive heat
left many of the soldiers, still wearing heavy uniforms, suffering
from heat exhaustion. The camels, dromedaries from the plains of
India, died in considerable numbers. Despite carrying quantities
of fodder for the animals, and food for the equivalent of a fair-sized
21st-century Scottish town, the traditional ability of the army to
forage for food en route proved impossible. The terrain through
which they were passing simply did not have enough food sources
available. Soon, the path of the army could easily be seen by the
quantities of animal remains left behind. It can only be imagined
that most of the commanders assumed this was an acceptable price
to pay for such a massive undertaking.

Events proceeded more or less according to plan; Kandahar
fell without a shot being fired. Ghazni, however, was felt to be
impregnable, reinforced as it was by 3,000 Afghans under the
command of one of the sons of Dost Mohammed. But on the night
of 22 July, one of the Bengali engineers managed to stack up huge
quantities of explosive by one of the main gates, which successfully
blew open the entrance. With the loss of just seventeen British
soldiers, but apparently hundreds of Afghans, the city fell to the
British within a day. By the time the army marched into Kabul,
they felt themselves to be wholly in charge of the situation and
the country. The cigars and polo ponies were about to come into
their own.

For Thomson, passing three months happily botanising, with
only some medical consultations causing minor interference, the
safety of the great walled town would have felt like a fortress, which
indeed it was. Contemporary illustrations depicting colossal walls
studded with tiny windows, semicircular fortified towers at all
strategic corners, and the figures of the officers dressed to the nines
in pressed and polished uniforms, gave credence to the feeling that

the British had arrived. For a few months they enjoyed an Indian summer, some officers sending for their wives, playing polo, and generally displaying all the characteristics of the British Raj. But while they entertained themselves, ignoring or misunderstanding local customs, the Afghan people were becoming incensed.

By the autumn, tensions were at boiling point. The beginning of the end was swift, although perhaps nobody could have predicted the ferocity of the bloodshed. In Kabul, Captain Alexander Burnes, a special envoy, was murdered. Alarmingly the 5,000 British troops and their families, plus 10,000 camp followers, were surrounded and outnumbered by Afghan forces. As the various high-ranking officers tried to negotiate their way out of the situation it was agreed with their Afghan captors that the entire contingent including women and children would be able to retreat safely in an easterly direction to Jalalabad, which was in the hands of the British Army. On 6 January 1842, as this contingent made the journey through narrow mountain passes, at the mercy of both snipers and the worst of the winter weather, it was difficult to know which was the more effective in decimating this increasingly desperate band. On 13 January, Dr William Brydon, an assistant army surgeon, staggered into view of Jalalabad, a city close to the Khyber Pass just under 100 miles from Kabul, mounted on a limping horse given to him by a dying Indian cavalryman and suffering from a gaping head wound. He was the sole survivor. Within one week, a massacre of staggering proportions had wiped out virtually 15,000 people.

Thomas Thomson had escaped being part of this massacre, as he was with a division at Ghazni, which had also been captured by the Afghans. He coped with his straitened circumstances in the one sure way he knew, academic study. He possessed just the two books. One was a Persian dictionary, and the other, allegedly, Lyell's *Principles of Geology*. There was one further avenue open to him. Despite having been in the area for a scant three months, he had learned the ways of the locals, and had observed that bribery was an essential part of life. He bribed one of the jailors to satisfying effect. The jailor agreed to teach him Persian. Falling captive in one of the many Afghan campaigns was bad enough, but things seemed about to get very much worse. The rumour was, as they marched north, that he and his fellow prisoners were destined for the slave market at Bokhara.

Eventually, though, all who had been held hostage at Ghazni were luckier than those at Kabul. They managed to negotiate their way out of the situation with bribery. Their release price was high: £2,000 and, as an insurance policy, they managed to persuade their captors to accept a pension for life from the British Government provided they would conduct them back to the advancing British Army. In fact an army division to relieve the situation was on its way, and it was probably they, and not the erstwhile hostages, who would more likely have been able to bribe the captors successfully.

Thomson had survived. Like most of his fellow men, he returned to India with nothing, no specimens, no notes on plants found, but his knowledge of Persian was greatly enhanced, and he clutched his dog-eared copies of his Persian dictionary and Lyell's *Geology*. He rejoined his regiment at Moradabad and was plunged straight into another bloody campaign. The following two years he was with the British regiments on the banks of the Indus River, and served throughout the Sutlej campaign.

Thomson might well have received such a bloodbath in India that he would have wanted to escape, if that were remotely possible, at the earliest opportunity. The army, though, had other ideas, and, whatever his sentiments, Thomas Thomson was to spend most of the remainder of his working life in India, apparently unable to release himself from the army, but his influential friend was working on his behalf. The long arm of William Hooker can be detected.

Thomson had been stationed in the foothills of the mountains at Lahore and Ferozepore until 1847, studying and collecting the local flora and engaging in such an educated correspondence that Hooker sought a more scientific appointment for him. Reading between the lines, Hooker endeavoured to find anything which would release him from being entrapped within the rigid hierarchy of army medical departments in remote areas of India, treating endless cases of various fevers, for which so little could be done. Thomson willingly found himself appointed as one of the three commissioners tasked to define precisely the boundary between China and Kashmir. While engaged on this gigantic task, the commissioners were to pursue and report on scientific interests, geology, archaeology and botany. Thomson was to concentrate on botany, geology and geography. This was by any standards an immense task, and fortunately he was not alone.

The party consisted of Thomson, Major Cunningham of the Royal Engineers and Captain Henry Strachey, who all set off from the hill station of Simla in August 1847, following the Sutlej Valley up to the Chinese frontier at Shipki. Here they were deflected from their most obvious route, as their opposite commissioners from China did not turn up and therefore they could not cross the boundary. The party then split up, each carving up a large section of mountain region to explore. Strachey left for the south-east towards Mansarowar Lake, and Cunningham for Kashmir. So Thomson was then alone, save for some local people, and his carefully copied references from an existing guidebook, written by William Moorcroft who had travelled there in 1820–21, to which he referred constantly. Thomson was also to publish a book on his experiences (*Travels in Ladakh 1820–21*), but not for many years.

Even today the area which he surveyed is sparsely covered by any guidebook, and for thirty years after his epic mapping, his own account stood firm. Thomson travelled in an area virtually unexplored – Moorcroft having traversed only a part of the area in which Thomson mapped. He wandered over areas of such remoteness that almost everything he reported was of interest. He was an excellent linguist, and so gleaned enough of the various languages to be able to communicate more than many of his contemporaries. Those hours teaching himself Persian when captured in Afghanistan paid off. He wove his way up and down the edges of the Zaskar Mountains on the edge of the great range which stretched east to present-day Nepal and Bhutan. The great mass of Tibet lay to the north, although he stayed mainly to the west of this area.

Although he was writing for himself, as well as for his employers, his account bears the civil servant's pen and style – detailed, detached, unemotional and, it has to be said, somewhat dry. He was no Robert Fortune, his comtemporary, who included so much of interest to the Victorian armchair adventurer. Although Thomson's narrative does not possess the 'wow' factor, he was a great gatherer of facts. High up in the Kussowlee Ridge, in the Himalayas, he observed that the fir *Pinus longifolia* is just like the Scots pine, but with very long leaves. Many were the temples he passed with large deodar trees – graceful, dense, conical conifers, with soft-textured leaves and bark – one of which was so large that 'its flattened trunk, as if formed by the union of two [grafted together] measured, at five feet from the base, thirty five and a half feet round. The grove was evidently of a great age.'

At 7,000 feet he was astonished to find not only staple foods, such as millet and buckwheat, but also walnuts, peaches, apricots and mulberries, all of which were 'common', while the vines were only just holding their own, the climate being not 'quite suitable'.

At Leh, the capital of the then province of Ladak, and the most important and only town in Western Tibet, about three miles from the Indus, he found the area precisely as when descibed in Moorcroft's mostly reliable *Travels*. Leh contained the principal monasteries of the area, and he wondered about the curious, long buildings, extending for more than half a mile, consisting of 'parallel walls, 12 or 15 feet apart, and nearly 6 feet high, the intervals between which are filled up with stones and rubbish, and the whole covered with a sloping roof, which rises to a gentle angle to the central roof ridge. On this roof are laid large slabs of slate every one of which is covered with Tibetan letters, or more rarely, with a rude drawing of a temple.'

By now it was late autumn, and in October they approached the Lazgung Pass, camping in temperatures of 15°F, and finding that water boiled at 184°F, which indicated that they were at 15,500 feet. He trekked up and down, meeting and purchasing some gold for a rupee from a man panning in the freezing-cold river, observing the boiling point of water at various points, and stocking up with tea and sugar; 'the brick tea [was] though not super excellent in quality, in the absence of better was quite good enough'. He watched incredulously the national game of 'Chaugan . . . which has been described as hockey on horseback, a definition so exact as to render a further detail unnecessary'. He observed the hunting of chakors, or partridges, and the glory of the Lake of Kashmir, covered with water lilies of at least three or four kinds, with the gardens of Shalimar rising up on terraces behind, pavilions of fine marble at the intersections of walks, and small canals. He found scarlet poppies and enchanting *Tulip stellata*, the pale creamy flowers of which open out in a flat star shape, and was all the time marvelling at the countryside of Kashmir, the 'sheets of cultivation, chiefly rice' and the cultivation of saffron, 'a very remunerative crop for the Raja, who retains the monopoly in his own lands, compelling the cultivators to sell the produce to him at a fixed price'.

Tibet had fascinated him, and also conveniently protected him from involvement in the second Sikh war. Finally he returned to Ferozepore where six months were allocated for him to write up his journals, and catalogue his collections. But his health was poor, and he was weakened by repeated attacks of fever. Eventually, recovering at Simla, he was all set for a return to England, when he happened to hear that his old Glaswegian schoolfriend Joseph Hooker had been travelling in the eastern Himalayas. He set out for Darjeeling to meet him. Hooker, however, had been imprisoned in Sikkim, and by the time he had escaped, and the two were reunited, Thomson had run out of time. Hooker was desperate that Thomson should travel with him for a year. Thomson took a gamble, and asked for a year's leave of absence without pay. For nearly two years the pair of them explored the Sikkim forests, until, with his health now wrecked, Thomson returned home. He received an unsympathetic reception from the East India Company who had little interest in the enormous variety of his collections, and even less in paying him to write about them or catalogue them. Thomson decided to publish and be dammed, even if he had to bear the expense himself.

He returned to India a couple of times. From 1854 until 1861 he was superintendent of the Botanic Gardens in Calcutta, in conjunction with the Professor of Botany at Calcutta Medical College. Afterwards, he returned to live at Kew as a 'confirmed invalid', although this did not prevent him travelling out as the secretary of an expedition to observe the eclipse in Northern India in 1871. It was a slow end to an energetic life, and he died just eight years later in London. Despite having struggled to be valued by the East India Company, he collected a raft of awards: he was a Fellow of the Linnean Society, the Royal Society and the Royal Geographical Society, from which he received their gold medal in 1866, and a founders' medal for 'his researches in Western Himalayas and Tibet'.

Underneath that rather dull-sounding exterior, Thomson had an eye for the flamboyant. Plants he brought back range from the *Cedrus deodara*, the graceful cedar subject to many a drawing and painting in its native setting, the tall pine trees from the same foothills of the Himalayas, and several primulas which have settled happily into British gardens; the *Primula rosea* and *P. denticulate* are ever popular. To Thomson is owed *Rhododendron arboreum* which soars 40 feet high, with red, white or pink flowers, *Viburnum grandiflorum* which scents the air at the end of winter through to spring, and the aptly named bulb, the Crown Imperial *Frittilaria* which makes a majestic show of orange or yellow flowers.

Thomson's rather tedious accounts of his travels have gathered dust over the years. Robert Fortune was to follow him eventually into the foothills of the Himalayas just a few years after his return in 1851. Unlike Thomson, however, Fortune had a firm financial agreement comfortably in his grasp.

CHAPTER 14

Robert Fortune (1813–1880)

FAME, FORTUNE AND CHINA TEA

———

Ten years almost to the day after David Douglas fell to his gory death in a Hawaiian wild bullock pit, a mere twig was all that separated Robert Fortune from a similar demise, or so he claimed. Unlike Douglas, Fortune, just thirty-one, was on the opposite side of the world, in China. And unlike Douglas, his luck held out. Ever the quick-thinking, physically powerful adventurer, he grabbed a branch – the 'twig' referred to above – to prevent himself landing headfirst in a waterlogged wild pig trap and drowning. Fortune was a gifted raconteur, and perhaps the tale gained a few embellishments in the telling.

He made his name in two spheres: botanical and literary. He found plants within China that lit up the eyes of generations of flower lovers. He plucked the tea industry from awkward trading partners, the Chinese, and bedded the entire industry down, complete with imported Chinese tea experts, in the northern hills of India. Then, finally, he polished off his rollicking life with a series of books which stood the test of time. He included much homely detail, understanding only too well that he could appeal to an audience who were not botanical zealots but were riveted by tales of tea with mandarins, descriptions of escapes from dastardly pirates and appreciative moments of admiration for beautiful Chinese women and ceramic arts. His cleverly amassed personal 'pension fund' included not only royalties from this successful writing formula, but also some of the very antiques about which he was eulogising. Their sale in Britain certainly boosted the Fortune coffers.

Fortune was born on 16 September 1812, just as Napoleon was massing his troops to dominate Russia. His background was modest. Born at Kelloe, Edrom, a backwater in the lowlands of Scotland, on leaving school he rapidly advanced to a gardening apprenticeship close to home, then moved over to Moredun, closer to Edinburgh, and finally into Edinburgh, to the Royal Botanic Garden, where he fell under the autocratic but influential head gardener William McNab. Not for the first time did McNab's influence pay dividends. Fortune used his time well in his two and a half years with McNab. Around 1840 it was McNab who ensured that Fortune was appointed superintendent of the hothouse department at the Horticultural Society's garden in London. Three years later, when Fortune was thirty years old, the Horticultural Society gave him a plum assignment on the other side of the world, collecting plants in China.

Robert Fortune grew into an opinionated, racy storyteller and a gutsy companion, his own self-importance tempered by an engaging self-deprecating sense of humour. He was a chameleon, a risk-taker, a crack shot, of substantial intelligence and a highly resourceful man. By the time his adventures got under way, he was able to play them out against a background of British imperialistic swagger, a British domination of world events, supported on the solid foundations of British superiority in world economics. Fortune was a calculating businessman, and fitted into the era to perfection.

China was ripe for picking up botanical wonder plants, especially as it was a country which had been extensively cultivated and – even if nineteenth-century Europe was reluctant to acknowledge it – sophisticated for thousands of years. While there were 'new' plants aplenty to extract from the wilder areas, many species, such as *Chrysanthemum*, had been nurtured and developed and were simply there for the successful barterer. The Chinese Opium Wars, in which the British had smashed their way with their superior armaments through the meagre Chinese defences, meant that Britain now had several toeholds around the coast of China. These coastal ports were cantons at Guangzhou, Xiamen (Amoy), Fuzhou (Foochow), Ningbo (Ningpo), and, of course, a certain rocky outcrop upon the future of which Fortune proffered a crushing aside:

> the island of Hong Kong has been ceded to England, the foreign population in it has been much changed. In former days there were only a few mercantile establishments, all known to each other, and generally most upright and honourable men. Now people from all countries, from England to Sydney, flock to the celestial country, and form a very motley group.
>
> Viewed as a place of trade, I fear Hong Kong will be a failure. The great export trade of southern China must necessarily be carried out at Canton, as heretofore, there being at present, at least, no inducement to bring back trade to Hong Kong. It will, nevertheless, be a place of great importance to many of the merchants, more particularly to those engaged in the opium trade; and will, in fact be the headquarters of all houses who have business on the coast, from the facilities of gaining early information regarding the state of the English and Indian markets, now the steamer communication has been established between this country and the south of China.

Fortune left little doubt that he found that China had been quite falsely described in the English press as being a type of 'fairyland' in a wondrous remote location, whose people were highly sophisticated, artistic, productive and scientific. He disagreed profoundly, feeling that when the local Chinese told him exactly what they thought he wanted to know, whether it was correct or not, it was not perhaps that this was symptomatic of their different customs, but rather that they were ignorant barbarians. In time his general opinion moved marginally. He was pragmatic enough to acknowledge that many of his new Chinese friends were good enough men, and that really 'China is just like other countries', implying that the mix of humanity included all types. It was also true enough, he acknowledged, that for centuries the Chinese had managed to make the most exquisite porcelain, lacquer ware and silks. Even more noteworthy was that they had been able to construct a compass for navigation. But what good had it done them? he wailed.

Grudgingly, Fortune did conclude that their agriculture was perhaps more civilised than that of other nations in the East, but not, of course, a patch on Europe, and Britain in particular. China, he sighed, had peaked many years ago and was now on a downward path. Fortune of course was writing these observations long after his first trip to China. Then, it was all new.

He arrived in China on 6 July 1843, as an envoy of the Horticultural Society after a four-month trip on a ship with the unlikely name of *Emu*. Not much had improved in the attitude of the Society's esteemed members towards generous reimbursements. They offered him for his personal protection a 'life preserver', which appeared to be a wooden club heavily weighted with lead, and £100. Fortune managed to extract a shotgun and a brace of pistols after some persuasion, on the strict proviso that he traded them back, promptly, on his return, reimbursing the Society for their purchase. His salary was no better than that of Francis Masson seventy years earlier, which had reached £100 a year by the time he retired. George Don who had gone to west Africa and the West Indies in 1821 had been paid £90 per year. This was not unusual in a British administration that sent generations of young men out along the spreading tentacles of the Colonial Service. Honour was expected to more than compensate for miserly pay, with accommodation and offices run on Scrooge-like principles. Despite the lack of pay and conditions, great things were expected of Fortune. The Society was giving credence to and pinning high hopes on a report from John Reeves, a retired tea inspector, who while working in Canton had been enchanted by the plants he had seen. Reeves had apparently been responsible for returning with *Wisteria sinensis*, Chinese wisteria, the most common one of all. Now he was an important member of the Horticultural Society's Chinese committee and it was only at his instigation that Fortune was going at all.

Fortune quickly discovered that the poverty and filth of the coastal ports he visited was quite appalling. The interior of China was still dangerous as any foreigner was fair game for mugging and murder. However, Fortune had to go up rivers and explore the interior as he had been charged with looking for a series of unlikely plants. The Society was anxious that he concentrated on bringing back hardy plants that would thrive outdoors in Britain, prove popular and bring a generous return. Therefore they thought that peonies and camellias would fit the bill nicely, as well as various groups such as citrus fruits – especially juicy satsumas – and some of the peaches which allegedly weighed about three pounds each.

The dice were loaded against Fortune: as a foreigner his appearance alone made him a walking target for robbers; he was being set down in a land with a language of which he had only rudimentary knowledge; and he understood but a smidgeon of the cultural life of even a tiny pocket of this vast country. He had his work cut out. On top of these difficulties, the Society wanted peonies that were blue, camellias that were yellow, and roses that were double-flowered and yellow. It was rather like being presented with a shopping list for run-of-the-mill products but with a bizarre twist: purple apples, jet black lettuce and scarlet milk.

Fortune rose to the challenge. Not only did he get to grips with the language, he adopted the Chinese way of dress in order to disguise himself, and he became a type of rogue trader, cajoling, bribing, and simply threatening anyone who would lead him to nurseries from which he could get the plants he wanted. And all the time he studied and commented on the Chinese, their food, language, laws (lacking), customs (strange), and many other subjects. On the role of Chinese woman he thundered, 'the females here are like those of the most of all barbarous nations, kept in the background, and are not considered as on an equality with their husbands. For example they never sit at the same table; and when a singsong or theatrical performance is got up, they are put in a place out of view, where they can see all that is going on and yet remain unseen.'

He traversed up and down the coast, from the much maligned Hong Kong all the way to Shanghai, aboard local boats, occasionally going up rivers and exploring alone. Most of these areas were within striking distance of British protection stationed at the various ports, and he noticed that all sorts of coins made up the currency, 'dollars, rupees, English shillings and sixpences, Dutch coins etc.'

Inland, there were mixed receptions. At one village, where he supposed that no foreigner had ever appeared before, the villagers were curious, offering him teacakes and a share of their bamboo pipes. However as he progressed closer to the main part of the village, he noticed that they became very suspicious, and tried to turn him back, trying every wile they could to deflect him from entering the hamlet. They pointed to the heavens where the black clouds threatened a storm, hoping to scare him off, but Fortune was set on carrying on. Finally, when all excuses seemed to fail, the villagers sent the children scuttling on ahead to sound an advance warning. Fortune, however, told of how he disarmed them immediately by agreeing to eat cakes and drink tea; they were immediately reassured by the evidence that he ate and drank exactly like themselves. Within a few minutes he was the centre of a curious crowd who searched for his pigtail, which to their astonishment was absent.

At every opportunity he absorbed a new lesson in Chinese customs and attitude, and there appeared a gradual softening of his attitude, even an occasional note of admiration creeping in. Tea with a senior government official, or mandarin, was taken with Rev. Abele, an American missionary in a suburb on Hong Kong Island. Having passed through various outer courts and politely sipped tea, the Reverend whispered to Fortune that much better tea would be available as soon as they reached the inner rooms with the mandarin himself. Good as it was to have the best quality tea, equally fascinating to Fortune was 'the great veneration with which the Chinese regard anything that is old. One of the pieces of porcelain, he informed us, has been in his family for 500 years, and had the peculiar property of preserving the flowers or fruits from decay for a lengthened period. He seemed to prize it as much on account of its age, and handled it with great veneration.' Fortune was also to take on board this 'veneration' of ancient items of china. He was to become a collector himself, and these items formed the basis of a collection which he was to sell at a fat profit when he finally returned to Britain. He was one of the few plant collectors who realised that curiosities from the countries they visited would be saleable in the UK.

Having explored the area around Amoy, Fortune wanted to set off for Ningpo and Shanghai. It was the first of a series of adventures. Sea travel was the main mode of transport. This time as they sailed past the Formosa Channel a storm blew up with such force that the waves threw up 'a large fish weighing at least 30 pounds which landed on the skylight on the poop, the frame of which was dashed to pieces, and the fish fell through, and landed upon the cabin table'. Fishing, as observed on a canal near Ningpo, was generally carried out in a far more genteel fashion, by the use of cormorants. Fortune was so fascinated by this when he first saw it that he sat in his boat with the sail dropped for hours entranced by the procedure.

On each boat was one man who had about ten or twelve birds, all of which were perched on the sides of the boat. At the command of the owner they scattered themselves over the canal searching for fish with their 'beautiful sea green eyes' and as quick as lightning diving on them. As soon as a bird surfaced with his trophy he was called back to the boat by the owner and obediently paddled to the side like a loyal dog. There, he disgorged his captured prey, which a ring round his neck had prevented him from swallowing. Sometimes the cormorants behaved like schoolboys truanting, rushing round playing and forgetting their duties. At that point the owner using the ever available bamboo stick would thrash the water and call out angrily, and the bird would immediately jump to attention.

Naturally Fortune wanted to acquire birds of his own. After lengthy discussions through the consulate in Shanghai he managed to find two pairs of birds and accompanied them on the boat from Shanghai to Hong Kong complete with their food, in this case a jar of eels in muddy fresh water. Dramatic as ever, a storm blew up and this time the cormorants on board added excitement. As Fortune put his head out on deck, the first thing he saw was the cormorants diving and screaming, and snatching eels which were sliding all over the deck. He realised immediately that the food intended to last until they got to Hong Kong was all either inside the birds, or consigned to the ocean. By the time they reached Hong Kong the birds had deteriorated to a poor condition. Never one to waste an opportunity, he killed them and preserved the skins. It was still an expensive loss, as the birds were highly valued and cost between thirty and forty shillings a pair.

In between travelling and sea adventures (and there were plenty more of those to come) Fortune spent his time scouring likely nurseries for plants. He knew that many of the plants his clients back home so avidly desired were already being carefully nurtured in China. It was laying his hands on them that was the problem. In many ways Fortune spent more of his time planning his next move, like a spy, than crawling over unexplored country. Not that his existence was any less dangerous and it certainly required nerves of steel, and diplomatic cunning.

First of all, Fortune had to find the nurseries, usually hidden behind high walls. This was challenging enough as the Chinese producers found it difficult to understand that he wished merely to trade with them. Either through jealousy or fear, they refused to divulge the location of their nurseries: 'If you want flowers', they said, 'there they are in the shops. Why do you not buy them? Shanghai men do this, and you should do the same.' If he asked directly for the plants, or tried to describe what he wanted, he was told that they would bring the plants to him, which, of course, never happened. Many excuses were offered such as that the nurseries were too far away. One day, returning from a walk with his servant he saw a new species of bird and shot it, the echoing bang bringing crowds of small boys to gaze in wonder at his gun. By demonstrating the gun he gained their confidence, and finally managed to persuade them to show him where the nursery was where he could buy flowers and plants. It was a breakthrough, and they conducted him to the very gates. He carefully noted the location but, as it was by then dark, he decided to return the next day.

His initial triumph was short-lived. He retraced his steps the following day, but the young lad rushed inside and in a trice the gates were slammed shut, and no amount of entreaties would persuade those inside to open up. Day after day went past and precisely the same thing happened. He approached at various times of day, from different directions, but to no avail.

Eventually he had to resort to the local British Consul, Captain Balfour. With one of the Chinese officers attached to the consulate to accompany Fortune, the two of them approached to exactly the same reception. The Chinese officer knew well that the gardening family would be hiding just behind the gate so he started to whisper to them, telling them how silly they were, and that the foreigner merely wanted to do business. Presently a rustling in the bushes indicated that they were changing their minds and eventually the door was opened. Not only did Fortune find the plants he had been searching for diligently all growing inside, well cared for, but it was by now late in the season so he was able to buy most of them in a dormant state and send them home.

That so many of Fortune's plants survived was due to his great faith and regular use of Wardian cases. These were glass cases, like miniature, airtight greenhouses, encased in wood and iron for strength, in which he packed his plants. They offered much the best method of plant transport and protection at the time.

The booty from the nursery that he had so doggedly pursued in the heart of China resulted in collections of tree peonies, Japanese anemones and Japanese cedars.

Often he discovered flowering treasures after death-defying escapes from the local people who regarded him as an infidel or foreign devil. After he almost ended his life in the wild boar pit, he found close by *Viburnum plicatum* v. *Sterile*, the Japanese snowball tree. As usual, his story is full of drama. Wandering near Canton he was stopped by a soldier whom he thought simply did not want him in the area – a request which, naturally, Fortune ignored. In a twinkling he was surrounded by aggressive groups of Chinese. Spotting the local walled cemetery he made for it swiftly, hoping to shake them off, but to no avail. Eventually he managed to stumble out, and race down the hill, being outmanned, almost outmanoeuvred and variously relieved of his hat, umbrella and coat. Just as safety appeared within sight, he was pelted by stones. Although badly winded, he eventually made it to the road and relative safety, finally arriving back at his lodgings with nothing more than sunstroke from the loss of his hat. He might not have brought back the snowball tree from this location, but he was to find it successfully later.

Much of the time it was the day-to-day occupation of the farmers round about him, allied with the customs of local people and everyday life, which were grist to his writing mill. Perhaps he was one of the first to describe the practice of 'green manuring', of planting crops, in this case clover, after the main crops were harvested, which are then ploughed into the ground to improve fertility. *Brassica chinensis* came in for detailed study too. Today this plant is commonly used in agriculture and known as oil seed rape. Fortune described how it was grown extensively, and noted that there was a great demand for the oil pressed from the seeds, which followed the brilliant yellow flowers. 'In April, when the fields are in bloom, the whole country seems tinged with gold, and the fragrance which fills the air, particularly after an April shower, is delightful.' Today, the cloying scent of the huge fields of oil seed rape do not attract the same effusive comments. Oil seed rape crops have become a common and profitable sight in many European countries.

While the agricultural practices fascinated him, it was the astonishing sight of golden azaleas clothing the hillsides on Chusan Island which made him hark back to memories of similar plants which were brought to the Chiswick fêtes. Beautiful as the Chiswick exhibits had been, they appeared as nothing compared with these vast ranges of flowering azaleas. A few azaleas had already reached London, and were carefully grown as precious rarities. They were not commonly available. For Fortune to see acres of azaleas was astonishing. Scrambling among them were clematis, wild roses, honeysuckles, glycerine plants (*Glycine sinensis*), and the Japanese camphor tree (*Laurus camphora*), the wood of which was made into chests, ideal for preventing clothes moths attacking voluminous Victorian clothes.

Not every method of cultivation in the area was to Fortune's liking. Dwarf trees rarely elicited his approval. He was ever astonished by the extent to which Chinese horticulturalists would go in order to produce curious and contorted plants. Not only would they spend inordinate amounts of time producing dwarfed trees, there was also a fascination for deliberately planting bulbs upside down so that the stems would grow in tortured fashions. Not every plant that the Chinese greatly valued elicited his approval either. Chrysanthemums, which were enormously popular, he found particularly distasteful. 'So high do these plants stand in the favour of the Chinese gardener, that he will cultivate them extensively, even against the wishes of his employer; and, in many instances he [the gardener] would rather leave his situation than give up the growth of his favourite flower.' He found he was not alone in his bewilderment at the time taken to grow the plants. He quoted a story of how he had been told of an old acquaintance who had grown chrysanthemums in his garden for no other purpose than to please his gardener, having no taste of the flowers himself.

In the meantime Fortune's search for the elusive yellow camellia was coming to naught even when he offered $10 to anybody who could produce even one. The reason was all too simple: camellias never have yellow flowers.

Less than a year after arriving in China he set off with the British Consul, a Mr Thom, and some other acquaintances to visit the green tea area of Ningpo, travelling in what he thought was a very civilised way in a mountain chair. This consisted of two long bamboo poles with a rudimentary chair positioned halfway, and was carried by two men at either end. He spent many months on the road and covered many hundreds of miles studying tea production. His greatest surprise was to find that, contrary to belief back in England, the black and green teas of the northern districts of China are both produced from the same variety of plant, *Thea viridis*, commonly called the green tea plant. He very closely observed exactly how tea was picked, dried, rolled and produced. He made detailed drawings, and described the process in different areas. His attention to detail regarding the entire process was to reap him many rewards. After he returned to England, he was approached by the East India Company with a grandiose idea: they wanted to establish, from scratch, a tea industry in the foothills of the Himalayas. In the late summer of 1848, he arrived back in China, disguised, as he had successfully been a few years before, as a Chinese person. He packed up over 20,000 plants and almost an equal number of smaller seedlings along with a clutch of Chinese experts and, by planting up this first mini-forest of tea plants in Assam and Sikkim, began the explosion of tea-drinking in Britain. By 1929 the trade had risen by over 800 per cent, and a new tea empire had been founded.

Almost nothing appeared to daunt Fortune. While still on his first adventures in China, in 1847, and well before he was charged with transplanting the tea industry, he decided to set off for Soo-chow, the Forbidden City, which was extremely dangerous for foreigners to enter. He hired a boat to take him upriver from Shanghai, where he shaved his head, acquiring a splendid secondhand pigtail 'of which some Chinaman in former days had doubtless been extremely vain', and reckoning that he made a 'pretty fair Chinaman'. As always, he employed a ruse, pretending to his Chinese servant and his boatmen that he was going but a short distance up the river from Shanghai, and only revealing the extent of his mission when well under way. He knew that bribery was also necessary. All was nearly lost soon after leaving when, on mooring at the first small town of Cading for a night, he was robbed of everything he owned – everything, that is, except for his cash, which he had prudently concealed under his pillow. The small matter of losing his clothes and possessions was no deflection. Purchasing new clothes, he reached the city, renowned for its culture and art, to find it lived up to his expectations, apart from a few dirty streets. The city, he found, was well guarded, the streets and lanes intersected with gates at various intervals, and closed for a curfew at nine or ten at night.

No one paid him the slightest attention. He was able to explore freely, gaze upon the ornamental lakes, and admire the women, noting that they lived up to their reputation as the prettiest in the country, being graceful and elegant, and dressed in considerable style in colourful, embroidered silks. The only faults he could find, although conceding that beauty was in the eye of the cultural beholder, were their deformed feet and chalky white faces. Although he found that the extent of the nurseries had been greatly exaggerated, and that this would not be a rich plant-buying trip, he did buy a double yellow rose and a white gardenia, which he proudly reported would soon be available the length and breadth of England. Back in Shanghai, he stepped ashore in full Chinese disguise, temporarily fooling his British friends.

Unlike so many plant collectors, Fortune rarely complained about the conditions in which he had to live, and his health held up for much longer than most, even in an area in which diseases appeared to wipe out locals and Europeans alike. He wrote many times of the mausoleums and Chinese customs of burial, sketching the sites, which were surrounded by pines. He noted the numerous and piteous graves of young children, with their little thatched roofs. Reports on death were never far from his pen.

Most dangers lay in fever and other illnesses and on visiting Hong Kong, he described it more in terms of health than geography:

> The island, particularly on the north-eastern and eastern sides, is very unhealthy; fever and cholera prevail to a great extent during the south-west monsoon and are almost always fatal. Our troops suffered far more from the climate, when they had possession of the place, than from the guns of the Chinese at the taking of Amoy. In the autumn of 1843 the sickness among the officers and men of the 18th Royal Irish was almost unprecedented, dismay was painted in every countenance, for every one had lost his comrades or his friend. It was dismal indeed. I have known many who were healthy and well one day, and on the morrow at sunset their remains were carried to their last resting place. The little English burial place was almost already nearly full, and the earth was red and fresh with recent interments. Scarcely a day passes without two or three being added to the number of the dead. And yet what was rather strange, a detachment of the 41st military, then commanded by Capt Hall, were, officers and men, all perfectly healthy; they were however, on a different part of the island.

Fortune had already embraced one Chinese habit that he agreed was less dangerous than baths in such a climate: the handing out of hot, damp towels before mealtimes in Chinese homes, a custom he found to be excellent. But his health did begin to crumble, and it could not have come at a more inauspicious time. Even he admitted a few years later that 'as long as I enjoyed health I got on well enough'. But he was feeling far from well on board his boat at the mouth of the Min River en route up the coast with his plant collections, having left Hong Kong destined for Shanghai.

It was by now the end of the summer of 1845, and Fortune was near the end of his three-year sojourn in the south of China. He was on board one of a large fleet of junks, totalling around 170. As he lay in his feverish bunk, negotiations were carrying on between the various captains all round about him, demanding that the mandarins send a convoy of warships to protect the fleet from pirates. Fortune was nonplussed: 'Nonsense! No pirates will attack us; and if they do, they will repent it.' He had no idea at that point of the extent of the piracy offshore, and put it all down to 'the cowardice of my informants'. The negotiations between the captains and the mandarins lasted for days. The storms blew from one direction, then from another. Fortune lay in his bunk lapsing in and out of consciousness with fever, and the ships went nowhere. He noted that the vessels 'never go to sea in stormy weather and, what with gales of the wind and negotiations with Mandarins, I was obliged to content myself with the junk life for a fortnight at the mouth of the river'.

He began to think mournfully that he would die on board before going anywhere, and that his grave would be on the banks of the River Min. Eventually the winds dropped, and the captain descended to Fortune's cabin asking for the umpteenth time if he had his gun and pistols in proper working order, and plenty of gunpowder. Fortune laughed it off, but did notice that the sailors were very uneasy about the voyage and would have been only too glad for more gales to keep them in harbour. To ready themselves, the Chinese sailors from every ship in the fleet prepared an offering to the gods:

> The tables were covered with dishes of pork, mutton, fruits, and vegetables. Candles and incense were burned upon the tables for a short time, and the whole business had something solemn and imposing about it. The Cook, who seemed to be the high priest, conducted all the ceremonies. On other days, as well as this it was part of his duty to light candles in the little temple where the gods were kept, as well as to burn incense and prostrate himself before them.

As soon as the fleet were out of sight of the Min River, with its spectacular scenery, they abandoned the agreed strategic plan whereby all the ships were to have kept together for mutual support, protection and safety in numbers, and carried on up the coast, making for a safe harbour at nightfall. Each ship progressed as fast as possible; clearly it was each man for himself. They were gone less than a day and were only fifty or sixty miles from the river, when the captain rushed down to inform Fortune, still lying on his bunk, that there were numbers of Jan-dous (pirate ships) waiting for them ahead. Ridiculous as Fortune thought this surmise might be, and convinced they were imagining every ship on the horizon to be a pirate ship, he still felt it might be prudent to prepare his gun. He found to his horror that when he trained his telescope on the ships ahead, all five of them, to be sure, were crowded with men.

Several thoughts rapidly came home to roost. Fortune realised that the ship's captain was considerably more intelligent than he had previously imagined; he realised that he, Fortune, might well be able to fight off one pirate boat, but hardly all five; and lastly he knew that he would be one of the first to be killed and thrown overboard, being a foreign devil. As he wrote in his diary, 'He devoutly wished himself anywhere rather than where he was.'

When he went up on deck, an even stranger scene was before his eyes, as the sailors were preparing for battle. It was a scene of utter confusion: money and valuables were being buried out of sight among the ballast, and baskets of small stones – the traditional armament – were brought up from the hold and put in caches round about the boat. Fortune realised that now all the pirates possessed guns and consequently 'a whole boat-load of stones could be of very little use against them'. As for the sailors, they, including Fortune's own servant, were now dressed in the worst sort of rags that they could find, their reasoning being that no pirate would think it worthwhile to kidnap a man in rags. Although they had every inch of sail up, and were going through the water as fast as they could, the crew were vanishing below decks to hide rather than manning the ship. Fortune brought out his pistols and ordered them up on deck. He judged them to be so cowardly that they would not even use the stones that they had so energetically brought up. In the meantime, the pirates were firing towards them, which allowed Fortune to judge the distance of their guns' capability.

He ordered the crew to take as much shelter as they possibly could within the boat, while he watched carefully. He saw the pirates – who were yelling and screaming, and generally trying to be as terrifying as possible – manoeuvring for the best possible firing position. When the pirates were within 20 yards of the boat, and were already splattering the sails with shot, Fortune raked the pirates' decks fore and aft with his double-barrelled gun, riddling the length of the boat with shot.

Had a thunderbolt fallen amongst them, they could not be more surprised. Doubtless many were wounded, and probably some killed. At all events, the whole of the crew, not fewer than 40 or 50 men, who in a moment before, crowded the deck, disappeared in a marvellous manner. They were so completely taken by surprise, that the junk was left without helmsman, her sails slacked off in the wind, and, as we were still carrying all sail and keeping on our right course, they were soon left a considerable way astern.

But it was not the end of the adventure. Pirate junk after pirate junk approached from different directions. Each time, Fortune employed the identical strategy, and finally, they all retreated and disappeared over the horizon. Fortune judged the danger to be over. Now that the pirates appeared to have been beaten off, the crew rushed on deck throwing stones after the retreating pirates yelling obscenities, and deriding them for not coming back to renew the fight, or so Fortune wrote. For the next short while he was held in high regard by the crew who knelt before him in gratitude. They proceeded up the coast from one safe anchorage to another. Fortune knew that their safety lay in arranging to sail in a convoy of small boats, but his crew were becoming overconfident in Fortune's ability to fight off the pirates. One evening, as they arrived and took their place at anchor alongside a number of junks in a small harbour, Fortune noticed that the other boats appeared to be making ready to sail, owing to the bright moonlight and high tide, to make progress up the coast to the next port.

Fortune was most anxious that they should proceed on, it being unwise to tarry and find themselves alone in the morning, but his entreaties fell on deaf ears. Less than an hour or so later they heard their neighbouring junks raise their anchors and set sail on the tide. Immediately they hauled up anchors, but had lost valuable time, and by the time they had made ready Fortune reckoned that the little armada must be several hours ahead of them. They set off, and only a short time elapsed when once more the captain came down to him in a state of considerable consternation and anxiety and pointing again at the horizon.

Fortune staggered above decks and trained his telescope on the ships in the distance – he was not nearly as sceptical as he had been before. The situation was indeed very dangerous and he was by no means convinced that he could keep the attackers at bay a second time. He deduced that the main armoury in his favour, apart from his meagre gunpower, was that the Chinese respected and feared foreigners on board because they imagined them to possess superior and numerous firearms. So he unpacked all his luggage, and ordered the crew to dress themselves up as Europeans. The short levers used for hoisting sails might, he hoped, fool the pirates into thinking they were firearms. Once more he waited for the enemy boats to come within range, which clearly required much sangfoid. It was all too much for his panic-stricken Chinese crew who rushed down below. Once more Fortune had to use his pistol to threaten his helmsman to stay at his post and, using exactly the same tactics as before, he managed to scare off the pirates.

That evening at a safe anchorage, with Fortune back in his bunk racked by fever, several other merchant junk ships limped into port, carrying severely wounded sailors, injured by gunfire. Fortune's crew pleaded with him to apply his medical knowledge to treat their wounds, but Fortune found that the iron shot used had caused appalling injuries, quite beyond his medical capabilities. The following morning when both wind and tide were favourable, he was hoping that they could leave promptly as they were now eighty miles from Chusan. This was one risk too many for his crew, who had been consulting between themselves and their neighbouring junks, and were intent upon waiting for the mandarins to supply armed protection.

When his request to sail was rejected, Fortune ordered his servant to go and search out another small boat, and informed the captain that he would abandon them, and take this new boat for his destination, Chusan. Such bluff had the desired effect: the crew pleaded with him to stay on board as their protector and reluctantly agreed to sail. In the meantime, the other junks asked to come in their wake, seeing in Fortune a saviour. Fortune refused, having overheard his crew giving an exaggerated version of the fights. When he woke up in the morning, he discovered that they were indeed anchored close to his destination, but the crew had changed their minds and wanted to detour to Ningpo first, before taking him, as agreed at the commencement of the trip, to Chusan. Fortune threatened them with his pistol, reminding them of the effect they had had on the pirate ships: 'Englishmen never allow promises which have been made to them to be broken with impunity. I know the way into Chusan Harbour as well as you do, and when the anchor is up I shall stand at the helm; and if the pilot attempts to steer to Ningpo, he must take the consequence.'

Finally, he managed to reach Shanghai, and a British doctor, under whose hands he regained his health. There was still one last treasure to pack up. As instructed by the Society, he had acquired some of the immense peaches, sometimes called Peking peaches, which rumour had it could be 12 inches in diameter. Fortune found some in the peach orchards south of Shanghai 12 ounces in weight, but not quite of the colossal size of legend. As he left Shanghai for Hong Kong, and eventually England, he gazed from his ship at the area around Shanghai where he had found the finest plants for his collections. Fortune arrived back in the Thames on 6 May 1846. By 20 October that year, he was writing the book which was to prove so popular. He also viewed his *Anemone japonica*, the waving pink and white flowers of which he had seen growing so frequently on the conical thatched roofs of Chinese graves, now growing just as luxuriantly in the Horticultural Society's gardens in Chiswick, London.

Fortune's fine journalistic eye was matched by an ability to write swiftly and accurately. As importantly, he had learned from the mistakes made by his plant-collecting predecessors, like Douglas and Menzies, who had taken years to write up their experiences, and missed a golden opportunity to turn their adventures into popular books. Within a few months he set to and brought out a book entitled *Three Years' Wandering in the Northern Provinces of China, including a Visit to the Tea, Silk and Cotton Countries; with an account of the Agriculture and Horticulture of the Chinese, New Plants etc.*, by Robert Fortune, botanical collector to the Horticultural Society of London, with illustrations. This appeared in 1847, published by John Murray of Albemarle Street, a considerable coup. The publishing house of John Murray was beginning to make a name for itself with guidebooks. Whereas so many of the plant collectors had been confined to and restricted by writing up their adventures under the censorship of the Horticultural Society, Fortune was taken under the wing of a professional publisher. Therefore his book was published with considerable literary and illustrative panache. The final page was of Fortune's drawing of a winding Chinese funeral procession, entitled *Finis*. Fortune does not appear to have disclosed the exact earnings from his books, but he certainly lived in comfort into old age, and his books contributed to his financial security.

However, this was by no means the end of Fortune's career as an adventurer and plant collector. He was by now, at thirty-six years old, experienced and nobody's fool. Least of all was he prepared to kowtow to the home-grown 'mandarins' of the Horticultural Society. He had switched allegiance: it was the East India Company who were anxious to establish the tea industry in the foothills of the Himalayas for the British government, and, allegedly, he was paid a whopping £5,000 for his part in this procedure. The East India Company were serious and successful entrepreneurs. Within the space of a decade or so, there would be teapots in nearly every house in Britain, from cotter's houses, almshouses and farmhouses, to mansions and Her Majesty's palaces. And all would be swilling with leaves plucked from descendants of these plants.

Fortune's reputation spread. The American government made plans to employ him to carry out a similar establishment of the tea plantations for their own use on their side of the Atlantic. But this was in 1858, and America had other important disruptions on its mind, as the unrest which led to the Civil War was already festering, and home-grown tea was less important than home-grown peace treaties.

Robert Fortune was to make one more trip to the East, this time to Japan, where hostility to foreigners was as vicious and murderous as it had been on his first trip to China. But Fortune managed to evade such problems, and gathered brilliantly coloured chrysanthemums. It is curious, perhaps, that his many drawings and descriptions of Chinese funerals should then be matched by his collections of chrysanthemums, a flower which he had initially despised, but which did become associated throughout Europe with funerals. However, things were far from doom and gloom for Fortune himself. Just before the publication of his book, he was appointed Curator of the Chelsea Physic Garden, and he visited Japan several times to collect chrysanthemums. He settled down for almost twenty years of happy retirement. He died in 1880, at the age of sixty-seven, and is buried at Edrom in Berwickshire.

Fortune appears to have been one of the most industrious of all the Scottish plant collectors, not only because he returned with such flowering exotica which possessed a huge advantage in that almost all could be grown in the British climate, but he also left very detailed travelogues, giving an insight into life in China, for instance, long before most Europeans managed to penetrate far inland. He was a man of many parts, and one of the few plant collectors who would transfer his experiences to the written word with ease and great success. He was a skilled artist and a man of huge courage. All year long, a Fortune plant is flowering in Britain, from his *Jasminum nudiflorum*, the yellow jasmine whose flowers spring out in winter before the leaves, often with snow and frost on the ground; then on to his *Primula japonica*, the candelabra primulas; *Forsythia viridissima*, one of the more delicate varieties; *Kerria japonica*, the yellow back-of-border plant which must appear in thousands of suburban gardens each year; the elegant, languid pink or white flowers of the *Camellia* of late spring; *Dicentra spectabilis* with its bobbing pink flowers, commonly known as Dutchmen's trousers or bleeding heart, in early summer; *Deutzia crenaat*, the summer-flowering white shrub; *Paeonia suffruticosa*, tree peonies with cupped pink flowers; *Skimmia japonica*, the red-berried upright shrub with perfect, sculptured leaves; *Wisteria sinensis*, its mauve flowers clothing even the tallest of house walls, and finally *Mahonia japonica*, a shrub with racemes of yellow flowers which appear from late autumn through to spring. Fortune's legacy is truly a rainbow of colour all through the year.

CHAPTER 15

Conclusion

There are many, many more tales of heroism and endurance still waiting to be told of the many more Scots who made their mark on the horticultural world, displaying just as much courage and fortitude even if only contributing a handful of plants to the great tapestry of botany from all over the world which is now commonly grown in Europe. This book only covers the major characters in the tale of the Scottish plant collectors.

While the nineteenth century drew to a close, with Europe's gardens burgeoning with plants from around the world, flurries of plants were still arriving in the sorting offices of those great gardening clearing houses of Kew, the national botanic gardens and flourishing nurseries such as those of Veitch and Dickson. By the turn of the century travel had become easier, and lifts seldom had to be hitched by negotiation with Her or His Majesty's Navy, the Hudson's Bay Company or the East India Company.

Today there has been a sea-change in attitudes towards collecting. Botanists of the twentieth and twenty-first centuries are just as dedicated and determined. But their remit has changed somewhat. Today's horticulturists travel less to tug up roots and bulbs for the delight and profit of the waiting gardeners of Europe, than to observe and save endangered species; a state of affairs foreseen by

George Don all those years ago in Angus, when he declared at the cusp between the eighteenth and nineteenth centuries that even some species of wild grasses in the Angus glens would be extinct ere long. This crusty old Scotsman, who died in such poverty, would be relieved to hear that more and more people are beginning to heed such warnings.

Today's botanists and horticulturists travel far to search out species that are on the edge of extinction. And how do they know that such plants exist in the wild? Because over the last few centuries the plant collectors roamed the world bringing plants back, and in addition to giving generations of gardeners great pleasure and profit, they inadvertently alerted us to their existence and contributed to the struggle to save these plants for our future.

Maps

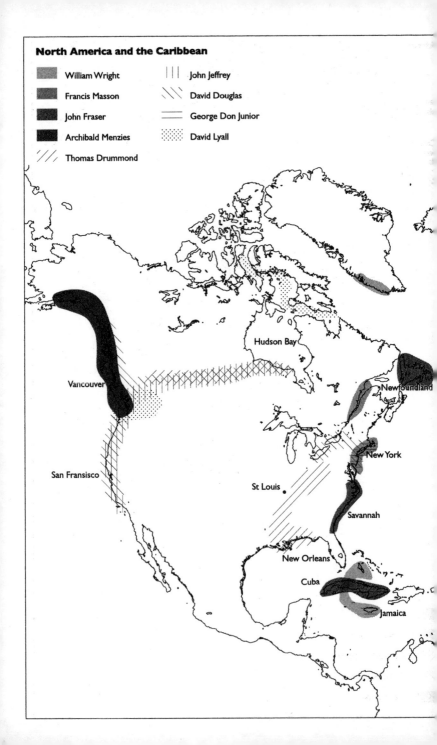

North America and the Caribbean

- William Wright
- Francis Masson
- John Fraser
- Archibald Menzies
- Thomas Drummond
- John Jeffrey
- David Douglas
- George Don Junior
- David Lyall

Vancouver

Hudson Bay

Newfoundland

New York

San Fransisco

St Louis

Savannah

New Orleans

Cuba

Jamaica

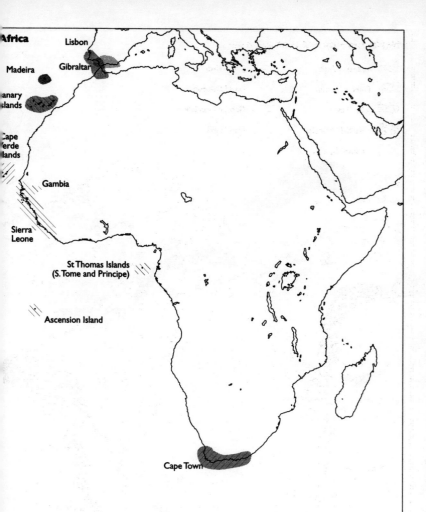

Africa

Lisbon

Madeira

Gibraltar

Canary
Islands

Cape
Verde
Islands

Gambia

Sierra
Leone

St Thomas Islands
(S. Tome and Principe)

Ascension Island

Cape Town

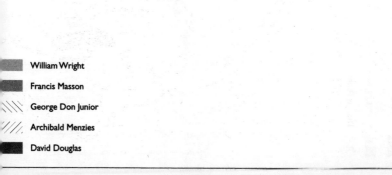

William Wright

Francis Masson

George Don Junior

Archibald Menzies

David Douglas

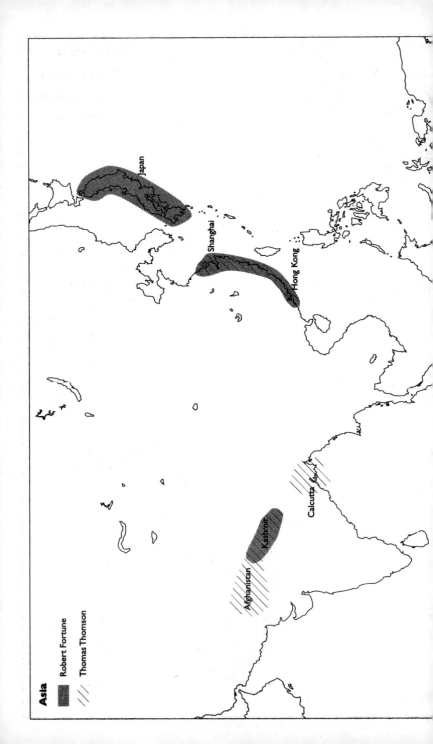

Asia

▓ Robert Fortune

/// Thomas Thomson

Japan

Shanghai

Hong Kong

Kashmir

Calcutta

Afghanistan

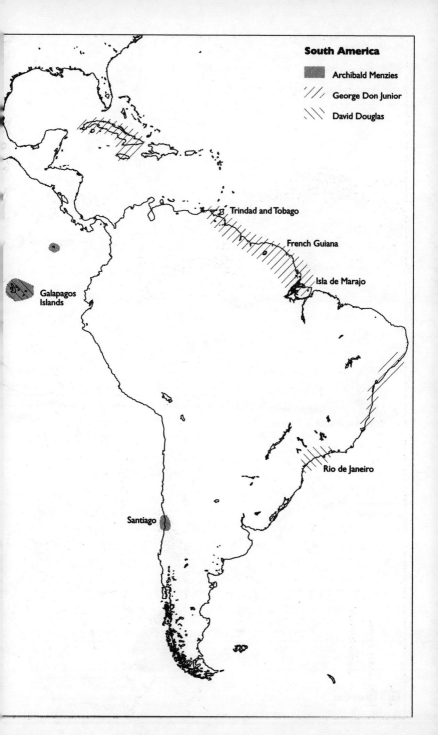

South America

- Archibald Menzies
- George Don Junior
- David Douglas

Trindad and Tobago

French Guiana

Isla de Marajo

Galapagos Islands

Rio de Janeiro

Santiago

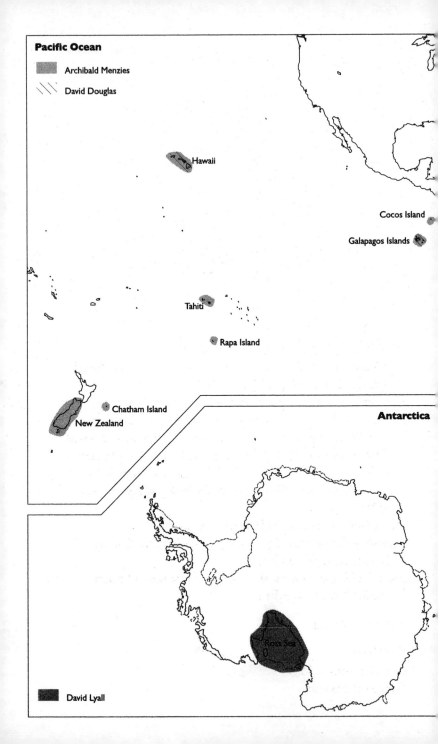

Pacific Ocean

Archibald Menzies

David Douglas

Hawaii

Cocos Island

Galapagos Islands

Tahiti

Rapa Island

Chatham Island

Antarctica

New Zealand

Ross Sea

David Lyall

Select Bibliography

MANUSCRIPT SOURCES

Bound Herbarium of William Wright's Hortus Jamaicense (Royal Botanic Garden, Edinburgh)

'Account of Three Journeys from the Cape Town into the Southern Parts of Africa; undertaken for the Discovery of New Plants, towards the Improvement of the Royal Botanical Gardens at Kew.' By Mr Francis Masson, one of his Majesty's Gardeners. Addressed to Sir John Pringle, Bart. P.R.S. in *Philosophical Transactions of the Royal Society of London*, vol 66 (1776)

1841 Census for the Parish of Clunie, Perthshire

Journals of George Don 1821–23 (Royal Horticultural Society, Lindley Library, London)

Journals of David Douglas 1827 (Royal Horticultural Society, Lindley Library, London).

PRINTED SOURCES

Periodicals

Curtis's Botanical Magazines
Gardeners' Chronicle

Journals of the Linnean Society
Proceedings of the Linnean Society
Proceedings of the Royal Geographical Society

Books and periodicals (specific issues)

Barthorp, Michael. *Afghan Wars and the North-West Frontier 1839–1947* (London 1982)

Bishop, George. *Travels in Imperial China: The Intrepid Explorations and Discoveries of Père Armand David* (London 1990)

Bourinot, John, ed. (1910) *Proceedings and Transactions of the Royal Society of Canada*

Coats, Alice M. *A Quest for Plants: A History of the Horticultural Explorers* (London 1969)

Coleman, E.C. *Captain Vancouver: North-West Navigator* (Whitby 1988)

The Cottage Gardener vol. 8 (1852)

The Deeside Field (Aberdeen 1920)

Desmond, Ray. *Dictionary of British and Irish Botanists and Horticulturists* (London 1994)

Douglas, David. *Journal kept by David Douglas During his Travels in North America 1823–1827* (London 1914)

Druce, G. Claridge. *Life and Work of George Don* (Botanical Society of Edinburgh 1905)

Fletcher, H.R. *The Story of the Royal Horticultural Society 1804–1968* (London 1969)

Fortune, Robert. *Three Years' Wandering in the Northern Provinces of China, including a Visit to the Tea, Silk and Cotton Countries; with an Account of the Agriculture and Horticulture of the Chinese, New Plants, etc.* (London, John Murray 1847)

Fortune, Robert. *A Journey to the Tea Countries of China: Sung-Io and the Bohea Hills; with a Short Notice of the East India Company's Tea Plantations in the Himilaya Mountains* (London, John Murray, 1852)

Fortune, Robert. *A Residence among the Chinese, Inland, on the Coast and at Sea, being a Narrative of Scenes and Adventures during a Third Visit to China from 1853 to 1856, including Notices of Many Natural Productions and Works of Art, the Culture of Silk, etc.* (London, John Murray, 1857)

Franklin, John. *Narrative of a Second Expedition to the Shores of the Polar Sea in the years 1825, 1826 and 1827* (London 1828)

Geyer-Kordesch, Johanna. History of the Royal College of Physicians and Surgeons of Glasgow: *Physicians and surgeons in Glasgow 1599–1858* (London 1999)

Grimshaw, Dr John. *The Gardener's Atlas* (London 1998)

Hooker, Sir William J. *Companion to the Botanical Magazine, Vol.2* (London 1836)

Hooker, Sir William J. *Curtis's Botanical Magazine, Biographical Sketch of John Fraser, the Botanical Collector born 1750–died 1811* (London 1836)

Johnstone, James Todd. *John Jeffrey and the Oregon Expedition. Notes from the Royal Botanic Garden (*Edinburgh 1939)

Johnson, J. *A guide for gentlemen studying medicine at the University of Edinburgh* (London 1792)

The Journal of Horticulture (1864)

Karsten, Mia C. 'Francis Masson: a Gardener-Botanist who collected at the Cape', in *Journal of South African Botany* (South Africa 1958 & 1959)

Landemann, George T. *Adventures and Recollections of Col. Geo. T. Landemann, late of the Corps of Royal Engineers, 2 vols* (London 1852)

Lange, Erwin F. 'John Jeffrey and the Oregon Botanical Expedition Oregon', in *Historical Quarterly* vol. 68, June 1967

Lindsay, Ann & House, Syd. *David Douglas, Explorer and Botanist* (London 1999 & 2005)

Le Rougetel, Hazel. *The Chelsea Gardener: Philip Miller 1691–1771* (London 1990)

Lynch, Michael. *Scotland: A New History* (London 1992)

McDiarmed, Rev. Mr James. *Statistical Account [of Scotland] of 1791, Parish of Weem*

Macintyre, Ben. *Josiah the Great: The True Story of the Man Who Would Be King* (London 2004)

McKelvey, Susan Delano. *Botanical Exploration of the Trans-Mississippi West 1790–1850* (USA 1955)

Mackenzie, Alexander. *Voyages from Montreal on the River St Lawrence through the Continent of North America to the Frozen and Pacific Oceans in the years 1789 and 1793, facsimile of 1801 edition* (Canada 1971)

Menzies, Archibald. *Menzies' Journal of Vancouver's Voyage, April to October 1792,* ed. Newcombe, C.F., MD (Canada 1923)

Millar, Rev. George. *Statistical Account of 1791, Parish of Clunie, Presbytery of Dunkeld, Synod of Perth and Stirling*

Minter, Sue. *The Apothecaries' Garden: The History of Chelsea Physic Garden* (Stroud, 2000)

Minute Book relating to the Oregon Expedition 1849–1859 (Royal Botanic Gardens, Edinburgh)

Musgrave, Toby, Gardner, Chris & Musgrave, Will. *The Plant Hunters: Two Hundred Years of Adventure and Discovery Around the World* (London 1998)

Nelson, Charles. 'James and Thomas Drummond: their Scottish origins and curatorships in Irish botanic gardens (*c.*1808–1831)', *Archives of Natural History* 1990 17 (1), 49–65.

Newman, Peter C. *Caesars of the Wilderness* (Canada 1987)

Nisbet, William. *The Edinburgh school of medicine . . . containing . . . anatomy . . . medical chemistry, and botany.*

O'Brian, Patrick. *Joseph Banks, a Life* (London 1987)

Ontario Historical Society. *Russell papers vol. 11, 1797–1798*

Paterson, Rev. W.P., DD. *History of Crieff* (Edinburgh & London 1912)

Kalm, Pehr. *Kalm's Account of his visit to England on his way to America in 1748. trans. Joseph Lucas* (London 1892)

Prebble, John. *Mutiny: Highland Regiments in Revolt 1743–1804* (London 2001)

Proceedings of the Linnean Society of London vol. 1, from November 1838 to June 1848 (London 1849)

Rembert, David H. jnr. *The Botanical Explorations of William Bartram in the Southeast*

Robertson, Forbes W. *Early Scottish Gardeners and their Plants 1650–1750* (Scotland 2000)

Roger, J. Grant. *George Don 1764–1814: The Scottish naturalist* (Edinburgh 1986)

Rosner, Lisa. *Medical education in the Age of Improvement: Edinburgh Students and Apprentices, 1760–1826* (Edinburgh c.1991)

Richard, Achille. *New elements of botany chiefly adapted for the use of students in medicine and pharmacy* (1829)

Shepherd, Sue. *Seeds of Fortune: A Gardening Dynasty* (London 2003)

Simpson, Marcus B. jnr, Moran, Stephen & Simpson, Sallie W. *Biographical Notes on John Fraser (1750–1811): plant nurseryman, explorer, and royal botanical collector to the Czar of Russia* (Archives of Natural History 1997)

Society for the History of Natural History. *Archives of Natural History 1936–1987 vol. 14* (Canada 1987)

Thomson, Leonard. *A History of South Africa* (USA 1995)

Thomson, Thomas. *Western Himalayas and Tibet: A Narrative of a Journey through the Mountains of Northern India during the Years 1847–48*

Wright, W. (Anon.) *Memoir of the late William Wright, M.D., Fellow of the Royal Societies of London and Edinburgh etc with Extracts from his correspondence and a selection of his papers on medical and botanical subjects* (London 1828)

Winch, N.J. *Correspondence of George Don 1764–1814; James Brodie of Brodie Castle, 1744–1824; James Edwin Smith, founder of the Linnean Society*

Index

An index to plants can be found following the main sequence

316 SEEDS OF BLOOD AND BEAUTY

INDEX TO PLANTS